MEDICINE
AN IMPERFECT SCIENCE

MEDICINE
AN IMPERFECT SCIENCE

EDITED BY NATASHA McENROE

SCALA

Produced exclusively for SCMG Enterprises Ltd
by Scala Arts & Heritage Publishers

First published in 2019 by
Scala Arts & Heritage Publishers Ltd
10 Lion Yard
Tremadoc Road
London SW4 7NQ, UK
www.scalapublishers.com

In association with
Science Museum
Exhibition Road
London SW7 2DD
www.sciencemuseum.org.uk

Every purchase supports the museum.

Editor: Charlotte Grievson
Design: Peter Dawson, Alice Kennedy-Owen at gradedesign.com

ISBN 978-1-78551-210-0

10 9 8 7 6 5 4 3 2 1

British Library Cataloguing in Publication Data.
A catalogue record for this book is available from the British Library.

Cover: Papier-mâché model of a human eye.
Science Museum Group. Object number 1996-277/11
Frontispiece: Pharmaceutical glassware in the Science Museum stores.
p. 7: Lancets and lancet cases from the Science Museum's medicine
collection.
p. 11: Bottle containing wormwood (*Artemisia absinthium*).

We are grateful to receive financial support
for this £24 million project from

TITLE FUNDER

PRINCIPAL SPONSOR

MAJOR FUNDER

The Wolfson* Foundation

MAJOR SPONSOR

Ω VITABIOTICS
SCIENCE OF HEALTHY LIVING

WITH ADDITIONAL SUPPORT FROM

Stavros Niarchos Foundation

CONTENTS

DIRECTOR'S FOREWORD

Medicine: The Wellcome Galleries form the biggest and most ambitious project that the Science Museum Group has undertaken in many decades. This permanent exhibition creates a centre for medicine that is of global importance, examining medical challenges of the past and the future through our world-class collections. In a series of five connecting galleries, our visitors can discover how we learn from the human body and how we influence the health of both individuals and communities. Intellectually stimulating and visually stunning, these galleries are relevant to us all.

The Medicine collections at the Science Museum are composed of two elements. The historic collection of Henry Wellcome, whose personal treasure trove has been on long-term loan to us for over 40 years, is complemented wonderfully by the Science Museum's own medical holdings. Often including cutting-edge medical technology, our active accumulation of material linked to scientific breakthroughs creates an ever-growing and ever-changing historic resource. We are delighted to work closely with the Wellcome Trust in our shared commitment to engaging a wide audience with bio-medicine issues both past and present.

We are grateful to our many funders of Medicine: The Wellcome Galleries, both organisations and individuals who have long supported the Science Museum and those who are newly involved. Their enthusiasm and support for this project has been truly inspiring. We would like to thank Wellcome (Title Funder), National Lottery Heritage Fund – and further support from GSK (Principal Sponsor), the Wolfson Foundation (Major Funder), Vitabiotics and the Lalvani family (Major Sponsor), with additional support from the Stavros Niarchos Foundation.

A project as vast as Medicine: The Wellcome Galleries has involved an enormous number of people, to whom we are extremely grateful for their time and expertise. Current healthcare practitioners, experts in the academic field and, perhaps most importantly, patients themselves have all shared their stories with us. At the Science Museum, one of our aims is to 'Ignite Curiosity' within all our visitors – something I am confident that we will achieve in this exceptional new space and through this book that acts as its companion.

SIR IAN BLATCHFORD
DIRECTOR OF THE SCIENCE MUSEUM GROUP

FUNDERS' FOREWORDS

WELLCOME TRUST

The pharmaceutical entrepreneur and philanthropist Sir Henry Wellcome – whose Will established the Wellcome Trust in 1936 – was also an avid collector. He sent his agents around the world in search of objects related to the ways in which people in different cultures have tried to preserve and restore their health over the centuries. His aim was to recreate humanity's medical past in order to study the roots of contemporary practices and inform future discoveries and inventions.

In his lifetime, Sir Henry accumulated more than a million items. Rich material for scholars, his collection was intended to inform, educate and entertain through public displays and exhibitions as well as research. It is gratifying, therefore, to see his collection at the heart of the new Medicine Galleries in the Science Museum.

Around 115,000 objects from Sir Henry's personal collection were placed on long-term loan to the Science Museum in the 1970s, and the links between our two organisations remain strong. With support from Wellcome, the Science Museum has been able to show many items from Sir Henry's collection over the past four decades, giving visitors different perspectives on the human experience of sickness and healing.

Wellcome's mission is to improve health. We fund not only researchers investigating the biology of health and disease, but also those exploring the cultural and social contexts of science and medicine. Understanding how ideas develop is essential to driving change and tackling health challenges. Equally important is that we engage and involve people with science and research.

The Medicine Galleries will help to ensure that Sir Henry's collection continues to inspire people. They offer new opportunities to find significance and meaning in the diverse stories of health and medicine that these objects represent.

BARONESS ELIZA MANNINGHAM-BULLER
CHAIR, WELLCOME TRUST

NATIONAL LOTTERY HERITAGE FUND

These fine galleries bring together three things in which the United Kingdom excels – medical innovation, curatorial skills and support for our heritage from the National Lottery. What the Science Museum has achieved, though, is far more than an engaging display of an unrivalled collection.

The new Medicine Galleries will be a centre for international consultation and research. Informed and inspired by a wealth of extraordinary artefacts and documents, those using the new galleries to these ends will be powerfully reminded of the importance of the improved health and wellbeing in modern society.

And it is entirely thanks to National Lottery players that the Fund I have the good fortune to chair has been able to provide almost £8 million towards these galleries, more than a third of the total project cost.

Indeed, since the Fund's inception 25 years ago we have invested more than £63 million in the Science Museum Group's sites around the UK.

Because our heritage is shared by all of us, and because our funds come from a wide cross-section of people, we are determined to support projects that reflect all aspects of that heritage and will have a wide appeal. This is exactly what the Medicine Galleries will do, with hands-on interactive technology that will appeal to people of all ages and will be fully accessible to individuals with disabilities.

The new galleries will provide a thought-provoking and fascinating experience for the visitors who will flock to them. My hope and expectation is that, as they explore the story of medicine from Roman times to the present day, many of those visitors will be inspired to take up careers in medicine and related sciences, or perhaps to become curators. But I hope all of those old enough to do so will be inspired to play the National Lottery and so help to fund more great projects like this one for many years to come.

SIR PETER LUFF
CHAIR OF THE NATIONAL LOTTERY HERITAGE FUND

GSK

At GSK, we are extremely proud of our long-term partnership with the Science Museum. For over 25 years we have worked alongside the Museum – from supporting the original Health Matters gallery in the early 1990s to becoming a principal supporter of the new Medicine Galleries. We have watched the Museum develop and expand, constantly looking forward whilst retaining its standing as a global centre of excellence, a place where generations of children and young people have first had their eyes opened to the wonder of science and its possibilities.

GSK, as a science-led global healthcare company, is fully committed to furthering science education, and looks to support programmes and projects that inspire every student to embrace the wonder and power of science, encouraging them to explore the amazing opportunities on offer. By bringing science to life, we aim to encourage the next generation of scientists and engineers because, without them, who will cure diseases and help solve the health challenges of the future?

It has been a privilege to be part of the Museum's exciting initiative to create a magnificent new home for its medicine collections based on Henry Wellcome's original collection. The Medicine Galleries are now literally at the heart of the Museum, giving a greater focus and understanding to the importance and relevance that medical development and advancement plays in all our lives.

We hope that you will enjoy your visit and this accompanying book will help to further your understanding and enjoyment of the galleries.

DR PAULINE WILLIAMS
SENIOR VICE PRESIDENT, GLOBAL HEALTH RESEARCH
& DEVELOPMENT, GSK

PREFACE

The Medicine collection at the Science Museum in London is one of the finest in the world. It is made up largely of Sir Henry Wellcome's personal collection, on long-term loan to the Science Museum since the 1970s. Wellcome, a pharmaceutical magnate, collected in the grand style of the Victorian age, believing that if he were to collect enough objects, it would be possible to calculate laws of humanity and learn the truth about our instinct for self-preservation. Since coming to the Science Museum, the collection has grown as curators have acquired objects relating to recent developments in health and medicine, adding to the historic material. Items range from a tiny Japanese netsuke carved from ivory to a large Victorian pharmacy shop. Endless rows of shelving in the Museum stores hold a multitude of blue and yellow ceramic jars once used to store drugs. There are countless drawers of surgical tools and whole rooms dedicated to X-rays and their associated ephemera, or to dentistry or fragile laboratory glassware. The collection includes groups of items brought together either for medical practice, or due to their association with a famous individual. There are teaching models and a whole range of curiosities linked to health. Caring for this vast and diverse collection takes a dedicated curatorial team who specialise in medicine and the role it plays in all our lives.

The role of the curator is not only to understand their collection but to interpret it for the public, to add to it, to publicise and protect it. A curator who works specifically with medicine will surely never struggle to present their collection as relevant – we each have our own history when it comes to medicine and are profoundly and enduringly invested in the health of ourselves and our loved ones. The personal connection that we have with the subjects of health and medicine adds an immediacy to the subject matter – but it also carries a potential risk unless it is communicated with sensitivity. Medical collections contain some of the most problematic objects in any museum collection – real human remains, for example, and images of diseased body parts, nudity or children's suffering. Historic collections such as that of Henry Wellcome also contain items collected during the colonial era that would have been acquired with none of the parameters set by current collection practices. Interpretation of all of these must be undertaken with great care and understanding.

This book is published on the occasion of the launch of a series of new galleries dedicated to showcasing the Medical collection at the Science Museum in South Kensington. Medicine: The Wellcome Galleries, which opened in 2019, form the most extensive display of the history of medicine in the world, telling the story of how we have managed our health at various points in time, and offering a broad perspective on medicine today. This book is not, however, a traditional catalogue of the galleries, nor does it seek to be an encyclopaedia of the history of medicine. Rather, it aims to express some of the interests and enthusiasms of the curators and historians who have created the new galleries, and whose professional research has been shaped by the collection itself. The authors of the book have acquired well over 100 years of curatorial experience between them and are united in their aim to share their passion for and knowledge of the history of medicine.

Medicine: An Imperfect Science is formed of a series of stand-alone, but connected, chapters through which we will learn of the history of the Medicine collection and how it came together, and what the objects themselves can tell us about this area of science. Some objects show us how scientists have gone about their research and practice, often through the way they are made or the materials they are formed of. Others can communicate something of the experience of patients of the past in a manner quite different from a written account of mental health, or other kinds of evidence. Analysing information from our bodies, in many ways the object at the centre of this particular history – whether physical measurements, imaging or dissecting the dead – has long been a tool for increasing our knowledge of health. The models and images that were used to support teaching and record medical information in the past are often visually striking and can straddle a boundary between record and portraiture. The objects within our collection at the Science Museum offer us a unique opportunity to engage with the history of medicine in previously unexplored ways. As well as allowing readers to learn more about the ever-growing collection of objects in the care of the Science Museum, this book provides behind-the-scenes glimpses of life in a large museum and the huge task of creating a suite of new galleries.

Each chapter of this book discusses, directly or indirectly, the problems inherent to medicine – whose status as a science can sometimes be called into question – and the telling of its stories. Medicine is an imperfect science and encompasses areas little understood today. The relationship of faith and trust between a patient and their practitioner and the support of loved ones have direct health outcomes, and even today, the immense power of the mind over the body is poorly understood. Through the historic Medicine collection at the Science Museum, we can reflect anew upon healthcare and our relationships to our own bodies.

NATASHA McENROE
KEEPER OF MEDICINE AT THE SCIENCE MUSEUM

WORMWOOD
Artemisia Absinthium L.
Compositae
Herb.   Europe   A674730.

1
COLLECTING MEDICINE
ROUTES AND ROOTS OF MEDICINE AT THE SCIENCE MUSEUM

SELINA HURLEY

The Medicine collection at the Science Museum is unique in its scale, geographical origin and scope. It touches on medical science, prosthetics, medications, deities from major global religions and relics from famous names in the history of medicine. The oldest objects in the collection date from 200 million years ago, with examples of dinosaur bones. These sit alongside the latest innovations in medicine that have been added more recently to the collection.

The collection has its beginnings in the extraordinary project of Henry Solomon Wellcome (1853–1936), a pharmaceutical businessman with an eye for advertising and a passion for collecting. His was a vast undertaking to collect the entire human experience of life and death, health and illness. Upon his death, his collection numbered over a million objects. A large part of this legacy was a library including anatomical treatises, recipe books, pamphlets, paintings and manuscripts, with at least 43 languages represented. The collection was put into the hands of the Wellcome Trust, created by Wellcome's will. The Wellcome Trustees decided to disperse the non-medical material to institutions across the world and through sale at auctions. The books and archives remain at the Wellcome Library. Exhibitions celebrating milestones of the history of medicine were still staged and some objects continued to be added, although at a much slower pace. The last and largest transfer of the holdings was announced in 1976 with objects to go on long-term loan to the Science Museum. A second, smaller transfer was completed in the mid-1980s. This meant a new home for the history of medicine and an upturn in collecting medicine at the Science Museum.

If each of the over 150,000 objects now in the care of the Science Museum could speak, revealing their histories, there would be a cacophony of voices.[1] For so many objects, the first evidence of their stories came with their entry into Wellcome's collection and they represent the work of hundreds of people over the collection's 140-year history. All the objects featured within this chapter now have their home at the Science Museum. For the first time, many of their routes to the Science Museum have been revealed using the original sources of the people who worked with them or used them.

Henry Wellcome (fig. 2) straddled two worlds at the turn of the 20th century – as one of the last gentleman collectors and one of the first pharmaceutical magnates.

Having grown up in Minnesota, Wellcome qualified as a pharmacist and started working as a travelling pharmaceutical salesman. In 1880 he was invited to join his friend, Silas Burroughs, who had emigrated to London, with licences to sell American products in the United Kingdom. Together, they founded Burroughs Wellcome & Co. Wellcome soon saw the opportunity to manufacture and sell their own products, and they set up their first factory in London. Their 'Tabloid' product range included compressed tablets, bandages, photographic chemicals, tea, medicine chests, and water testing and purifying equipment.

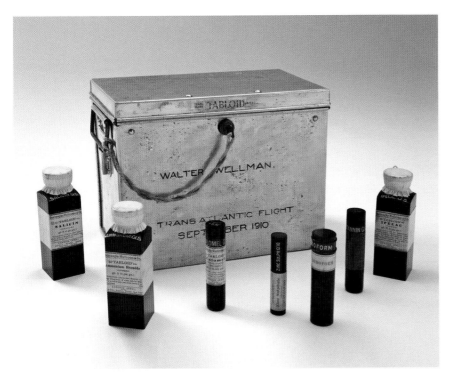

Fig. 2 (opposite)
Henry Wellcome, portrait
photograph, 1878.

Fig. 3 (above)
Wellmann and Vaniman's
medicine chest for their
attempt at transatlantic
flight in 1910.

Science Museum Group.
Object number A700006

Fig. 4 (above right)
Walter Wellmann.

The company became one of the first multinational pharmaceutical companies with outposts and research laboratories around the world. Today, the company is part of GSK (formerly GlaxoSmithKline).[2]

'Tabloid' was the brand that Burroughs Wellcome & Co. created as a memorable name for their products and, more importantly, one that could be trademarked. In 1884 Wellcome invented the word Tabloid, '"a pat name", a "fancy" meaningless word' to describe the company's new compressed tablets.[3] The single-dose compressed tablet was unheard of in the United Kingdom at the time.[4] The Tabloid name was applied to a range of products. Burroughs Wellcome & Co. donated medicine chests containing bandages, painkillers and fever and burns treatments to prominent explorers, adventurers and medical men of the day in return for their endorsement of the Tabloid products.

One such chest (fig. 3) was used by the explorer and journalist Walter Wellmann (fig. 4) and the engineer Melvin Vaniman during their attempt to cross the Atlantic Ocean in the powered airship *America* in September 1910. Sadly, their journey was unsuccessful and Vaniman died on a test flight for their second attempt in 1912, when the airship, *Akron*, exploded.

Wellmann's comments about the chest were immortalised in a book published by Burroughs Wellcome & Co. for the *Century of Progress* exposition in Chicago in 1934:

'Tabloid' Medical Equipment was the only one carried in the airship 'America' during one thousand miles' flight over the Atlantic Ocean. We had several occasions to use its contents for minor troubles, and found it completely and wholly satisfactory, which was but repeating the experiences I have had with your equipment on my expeditions to the Arctic regions.[5]

Chests that had been taken on expeditions were often donated back to the company to be used for publicity and, frequently, they ended up fuelling Wellcome's other passion – collecting.

Wellcome was deeply interested in the material culture relating to people's experience of health and illness, birth and death, faith and science, across the globe and across time. He had arrived in London in 1880 with a small collection of objects, which he displayed in his home.[6] What set Wellcome apart from other gentleman collectors was his ambition to 'collect the world'.[7] With a growing income from his success in the pharmaceutical business, he set out to achieve this ambition. While the origins of Wellcome's collecting are difficult to trace, the historian Frances Larson's excellent book *An Infinity of Things: How Sir*

Henry Wellcome Collected the World reveals that some of his early collecting was inspired by his friendships with Native American communities in his home state of Minnesota.[8]

Of the many thousands of objects now at the Science Museum, we know of the few objects that are recorded as having been collected by Wellcome personally rather than by one of his many employees. Apt within his scheme to collect the whole human experience of health, one object has its place at the beginning of the life cycle. It is a baby carrier of the Sioux people (fig. 5), accompanied by a handwritten note (fig. 6): 'Museum label says "Collected by Sir Henry Wellcome in U.S.A."'.

It is likely that the model baby now considered part of the object was added later for an exhibition, *Child Welfare through the Ages* (fig. 7), at the Wellcome Historical Medical Museum in 1954. There is no evidence to suggest that it was ever displayed at any other time at Wellcome's museum. As part of Wellcome: The Medicine Galleries at the Science Museum, it will once again be enjoyed by visitors for years to come.

However, we have no clue as to how Wellcome came across the baby carrier, who gave it to him, or what the acquisition negotiations may have been. There is no evidence that Wellcome saw or interacted with this individual item or expressed how he felt about it. It could simply be that the act of adding the object to his collection was enough for him.

In his burgeoning collection, Wellcome envisaged a three-dimensional encyclopaedia of objects, organised chronologically into displays (figs 8 and 9). This was brought to reality by an army of staff working behind the scenes. Versions of his museum graced premises at Wigmore Street in London from 1913 to 1920, and there were smaller-scale displays at Euston Road from 1936 to the late 1970s.

Amassing such a large collection was of course beyond the capacity of one man alone. Wellcome had several agents collecting on his behalf, all over the world, each of whom reported rigorously to him via letter or telegram, keeping him and his senior museum staff apprised of their movements. Agents were given funds to spend on the collection, but most of the purchases needed prior approval from Wellcome.[9]

Fig. 5
Baby carrier, with later added model baby, Sioux people, USA, 1880–1920.

Science Museum Group. Object number A655883

Fig. 6
Museum record for the provenance of the Sioux baby carrier.

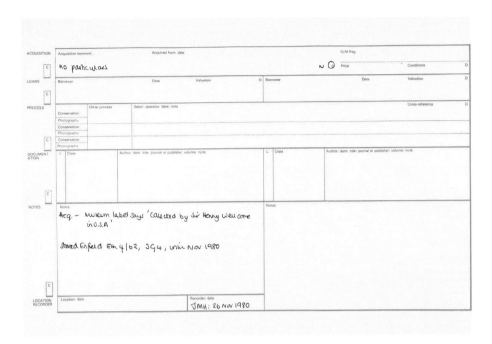

Fig. 7
Display from *Child Welfare through the Ages* exhibition at the Wellcome Historical Medical Museum, 1954, showing the Sioux baby carrier amongst other items for carrying infants.

Figs 8 and 9
Views of the Wellcome
Historical Medical Museum
from the 1950s (above)
and 1927 (below), showing
microscopes, Greek
pottery, paintings,
statues, pharmacy jars
and material relating to
the First World War.

One agent who enjoyed much more freedom than the others was Captain Peter Johnston-Saint (1886–1974; fig. 10). A Cambridge University history graduate in 1907, Johnston-Saint served as an officer in the Indian Army and Royal Flying Corps, before joining Wellcome in 1920 as Secretary of the collection. His impressive skill in securing high-profile acquisitions from 1926 onwards saw him elevated in 1928 to the role of ambassador for Wellcome's collection abroad. In 1934 he became Conservator of the entire collection. He was suave, well connected and persuasive; his diaries, now held at the Wellcome Library, contain intricate details of the thousands of miles he travelled on behalf of Wellcome.

On hearing the news that the wards at the Glasgow Royal Infirmary were being demolished in 1926, the hospital where the surgeon Joseph Lister (1827–1912) had developed his system of antisepsis, Johnston-Saint was dispatched to see what he could procure. The list covers everything from the mundane to the specifically medical – from a door handle to the ward that Lister would have used daily, to an air ventilator from one of the wards.

Johnston-Saint reported:

On my arrival in Glasgow on Monday, I visited Christie, the Contractor, and made arrangements for all the material available at the Royal Infirmary … I then saw Mr Andrew Brown, who made the original tin receptacles for Lister in the Infirmary in the eighteen-sixties. He has three steam sprays of different types, one of which is the first steam spray which Lister actually used in Glasgow. He has promised to let us have these on loan, and we shall probably have them indefinitely as he is very interested in all we are doing in the matter of the Lister memorial.[10]

Johnston-Saint makes it seem simple, but one can imagine the practical problems of gaining entry to a demolished site and physically removing items, before transporting everything back to London. Maker Andrew Brown (1852–1945) donated a carbolic spray and also made a replica of one of the versions known as the donkey engine (fig. 11).

Fig. 10
Captain Johnston-Saint, 1934–47.

Fig. 11
Replica of a donkey engine made by Andrew Brown to Joseph Lister's original design.

Science Museum Group. Object number A55244

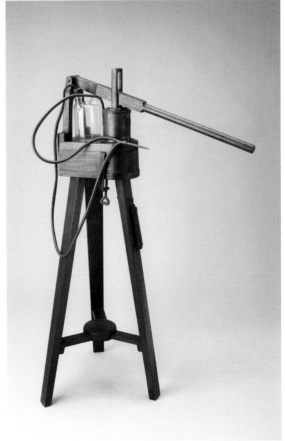

The original donkey engine was used around 1871 while conducting surgery or dressing wounds. It was designed to cover everyone and everything with a fine spray of carbolic acid, creating an antiseptic environment. Brown also made copies of a range of items such as milk jugs, cups, spittoons, patient record holders and candlesticks (figs 12 and 13). He donated these as well as the templates for them to Wellcome's collection.

It was not an uncommon practice for Wellcome to acquire copies or casts where the real thing was unavailable, or its owner was unwilling to sell or donate the original. While the replicas have become part of the story of the Wellcome collection, today the focus has shifted from completing a set by acquiring a copy to acquiring the object and its story.

Andrew Brown's family got in touch with the Science Museum 90 years on, and this unexpected contact allowed us to build a more complete record of his life and his workshop (figs 14 and 15). We can now match a face to the name in Johnston-Saint's records. Research on items held in the collection is ongoing and there is no telling where new information may come from or what untold stories remain to be uncovered.

By taking ownership of the material from Glasgow, Wellcome's collection had effectively commandeered Lister's legacy and historical reputation. It led to the Wellcome Historical Medical Museum becoming a focus for the centenary celebrations of Lister's birth in 1927, with an exhibition showing many of the items Johnston-Saint collected.

Johnston-Saint's work and that of Wellcome's collection did not go unnoticed. A letter from a member of the Glasgow City Chambers, whom Johnston-Saint had met on his visit, remarked: 'they [Glasgow Royal Infirmary] are now realising the mistake they made when they removed the Lister ward ... You are doing yeoman work and will have the great honour of having the most complete museum in existence of Lister'.[11]

Lister was not the only well-known name from this period of history to be represented extensively in the collection. Johnston-Saint also collected material from the pioneer of

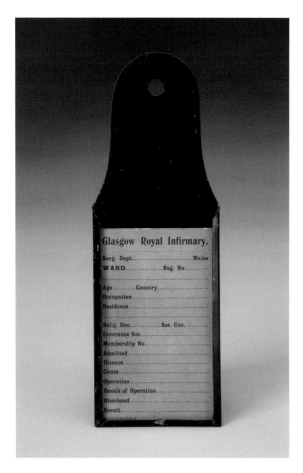

ANDREW BROWN,

FORMERLY WITH & SUCCESSOR TO

—(J. McVICAR & SON, ESTABLISHED 1809.)—

MAKER of ASEPTIC FURNITURE and SURGICAL APPLIANCES for HOSPITALS.

Tinsmith, Gasfitter, Heating and Electrical Engineer.

104, GEORGE STREET,

OPPOSITE N. ALBION ST.

GLASGOW

TELEPHONE:

BELL No. 118.

microbiology, Louis Pasteur (1822–1895), and that of vaccination, Edward Jenner (1749–1823), both key figures in the history of medicine. Some biographers of Wellcome comment that his collection 'helped secure his acceptance by the world of medicine and science, to which he had no professional entry'.[12] However, he was a qualified pharmacist, building a massive and successful pharmaceutical business. One way of looking at Wellcome's collection is to see it as his attempt to position himself in a lineage of scientific heritage, with Burroughs Wellcome & Co. products and his own research laboratories contributing to the state of medicine in his time.

Johnston-Saint was able to use his contacts in the higher echelons of society to secure some particularly extraordinary objects. As a child, he had visited Balmoral and played with the grandchildren of Queen Victoria. Years later, one of those children, Princess Victoria Eugenie of Battenberg, became the Queen of Spain. This childhood connection enabled him to gain an audience with the queen and King Alphonso XIII in 1930. There, he acquired a series of hide-covered bags, known as 'serons', from the first ever European expedition to Peru in 1777, home to cinchona plants (figs 16 and 17). Each seron contains cinchona bark, which contains quinine, and was used to reduce fevers and treat malaria. The serons were shown at the 1930 Wellcome Historical Medical Museum exhibition to commemorate the first European use of cinchona.

Since Wellcome headed up Burroughs Wellcome & Co., producing quinine preparations was extremely important – the plant's active part was a huge portion of the business. The ingredient found its way into every Tabloid medicine chest as well as into cold and flu remedies. As a young pharmacist, Wellcome travelled to Peru and Ecuador in 1878 to seek out the original cinchona bark sources.

Johnston-Saint also collected whole pharmacies as well as individual items. On 29 January 1932, he returned to a shop in Sharia el Marrakh, Egypt, that he had come across two years previously. At that time, he had asked the owner 'to keep a look out for a really fine Arabian drug shop'.[13] In the intervening years, the shop owner's son had been to Damascus and found such an old shop:

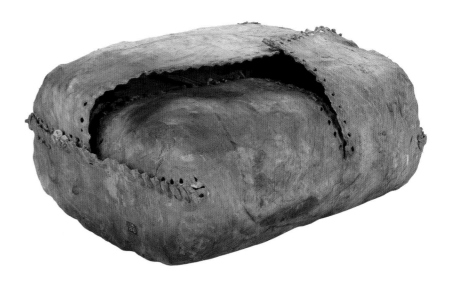

Fig. 16 (above)
Telegram sent by Johnston-Saint from Madrid, informing Wellcome Historical Medical Museum of the acquisition of serons from King Alphonso XIII of Spain.

Fig. 17 (below)
One of the serons acquired by Johnston-Saint filled with cinchona bark, from one of the European expeditions to Peru carried out by Spanish botanists Hipólito Ruiz and José Antonio Pavón in 1777–88.

Science Museum Group. Object number A654763

at least 16th century, probably 15th and brought most of it down to Cairo including a carved and coloured entrance arch (18th century), a ceiling in five sections, inlaid mosaic murabs or arches, three back windows, a 'curious counter, inlaid with mother of pearl with four drawers ... and a small slot into which the money was dropped, large faience drug pots, mortars, pestles, lamps and amulets'.

Johnston-Saint raved about one of the doors as a 'treasure ... beautifully inlaid with large bosses made of blue enamel on copper – a real 15th century piece'.[14]

However, despite the wealth of material gathered, Johnston-Saint discovered that most of the pharmacy pots had already been bought by a merchant in Cairo. The seller must have had too good an offer to refuse. Still, Johnston-Saint tracked the merchant down and offered him £45 for a total of 46 pots, including 34 with a blue-green glaze featuring designs of plants, birds, fishes and human faces (fig. 18). The remainder of the pharmacy cost £100. He had

a great deal of trust in the shop owner with regards to the provenance of the objects, where a gentleman's agreement appears to have been enough. In turn the shop owner had had the confidence that Johnston-Saint would return to purchase the items. Many of the acquisitions made by Wellcome's agents relied on personal relationships and contacts cultivated through meeting, talking and sometimes gently persuading people over a long period of time to donate or to sell objects.

Although he had no formal training in anthropology or archaeology, like many of his Wellcome Historical Medicine Museum colleagues, Johnston-Saint felt confident enough to date the pharmacy and its contents. He also saw its potential use: 'If we reconstructed this shop in our new building we shall have the finest Arabian druggist shop in any museum in the world. I have poked around a good deal in the course of my travels, but I have never seen such material.'[15]

Pharmacy reconstructions were an important method of presentation at the Wellcome Historical Medical Museum. A version of the Arab pharmacy (figs 19–22) remained on display from its acquistion until it was deconstructed and transferred to the Science Museum in the early 1980s as the last part of the transfer of objects. It was one of six different reconstructed pharmacies from different times and places. The importance of reconstructions to Wellcome's schemes, which reflected his work in pharmaceuticals, probably ensured their survival. Today, the components of the Arab pharmacy are in storage, but the pharmacy pots will be part of the new Medicine Galleries, on show for the first time in almost 40 years.

As well as his travels abroad, Johnston-Saint also collected closer to home. On the death of his father-in-law, the physician and geologist William Mansell MacCulloch (1849–1924), Johnston-Saint's wife donated her father's instruments to Wellcome's collection. These included dental descalers, a tongue scraper, two ophthalmoscopes, surgical instruments, a university club badge from 1864, an inhaler, two shell amulets from Africa, a glass pulsometer, a Spanish pottery feeding bottle and a leather-bound notebook used for note-taking at Professor Thomas Henry Huxley's biology lectures in 1880, all of which remain at the Science Museum. At the time, it was promised that they would be kept 'as near as possible together' and join the 'relics of distinguished physicians who have passed away'.[16] The broad range of items suggests that MacCulloch was also something of a gentleman collector, albeit on a smaller scale than Wellcome. As with Johnston-Saint's wife, it was common for people to donate items on the death of a relative who had worked in medicine to Wellcome's collection.

Johnston-Saint served at Wellcome for 27 years before retiring in 1947. During his career there, he collected thousands of objects and received numerous honours, including La Croix de Chevalier de la Légion d'Honneur alongside Henry Wellcome in 1934. This was presented to them for their work on the history of medicine in France.

Not all agents enjoyed the same freedoms as Johnston-Saint. Catherine Georgievsky, a Russian émigrée (1898–1944), had a more tightly controlled project both in terms of geography and expenditure. How she came to work at Wellcome remains a mystery, but we do know that she became a naturalised British citizen in 1931. The geographical focus for her collecting was

Czechoslovakia (as it was then known), sometimes working with her brother, John, to secure items for the museum. Her work focused on the healing waters and spas of the region, the local landscape of chemist shops and the history of religious orders providing medical care.

During one of many trips to Prague, in 1933 Georgievsky met with a monk from the Milosrdni Bratri Monastery and Hospital. While undertaking a tour, she spotted a glass jar marked *Pulvis Chinae F. C.* – Latin for 'powdered cinchona' – from the monastery's 1627 pharmacy.

Acquiring the jar from the Milosrdni Bratri Monastery and Hospital was difficult, and Georgievsky was told on numerous occasions that it was unavailable. She found out that the monk with whom she was liaising was keen on postage stamps and she took some to offer as a gift on her next visit. Though she did not manage to purchase the jar itself on that occasion, the stamps were received gratefully enough that Georgievsky was able to procure photographs of it, one of which survives today in the Wellcome Library (fig. 23).

In a futher attempt to gain the monk's favour, Georgivesky gifted to him a glass tumbler produced to commemorate the first issue of stamps from the Austrian State Post Office:

On my last visit to the Monastery I took the tumbler with me and showed it to the monk drawing his attention to the perfect reproduction of stamps and saying that it had also belonged to a chemist shop. I suggested an exchange, his face brightened up … He rang for the head chemist and told him to bring the keys of the old chemist shop and we all went up. Here he consulted the chemist first if he should part with the jar, the [chemist] only shrugged his shoulders and coming close to me asked me – how did you find out his weakness? The general opinion was that if the jar was of interest to our Museum it will be appreciated and therefore we were to have it … I was very pleased to have succeeded in getting it as the Conservator's [Johnston-Saint] instructions were to do my very best to secure it.[17]

It remains one of the earliest pharmacy jars to have held cinchona powder within the Science Museum's collection today (fig. 24), all thanks to the persuasiveness and individual efforts of

Fig. 19
A version of the Arab pharmacy on display at the Wellcome Historical Medical Museum containing many of the items that Johnston-Saint collected.

Fig. 20
Wooden cupboard with inlaid tortoiseshell and mother-of-pearl from the Arab pharmacy.
Science Museum Group. Object number A643477

Fig. 21
Glass lamp with painted decoration.
Science Museum Group. Object number A643433

Fig. 22
Brass and copper jug for rose water.
Science Museum Group. Object number A123065

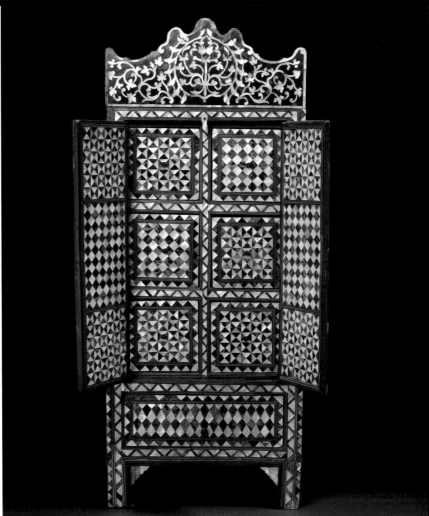

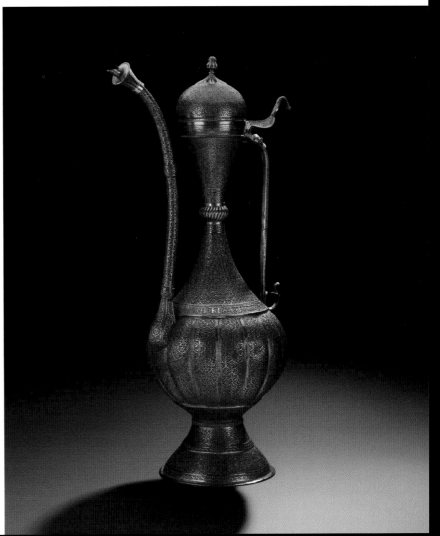

Fig. 23
Photograph of pharmacy jar from Georgievsky's trip to Milosrdni Bratri Monastery and Hospital, 1933.

Fig. 24
Jar originally containing cinchona, 1627. Obtained by Georgivesky for the Wellcome Collection.

Science Museum Group. Object number A137858

Catherine Georgievsky. After her many collecting trips were over, she worked in the Autograph Department of Wellcome's collection. It consists mainly of items that have been signed by identifiable individuals, famed within the history of medicine. Until her sudden death from septicaemia in March 1944, she described, repaired and filed thousands of letters signed by people such as Samuel Pepys, various polar explorers of the day and Pierre and Marie Curie, the discoverers of radioactivity.

Not all acquisitions required this level of relationship building. Purchasing lots at auction was another route of Wellcome's collecting. He sent decoys to auctions in order to conceal what types of object he was interested in, so as to get them as cheaply as possible. Often two decoys would be at the same auctions. Auctions were attended at least three times a week, depending on what the sales catalogues promised.

One such decoy was Frank Webb. His day-to-day job was as chief technician in charge of fixing displays and general maintenance of the museum, but he often found himself on the auction floor with strict instructions to bid for

particular items. At one auction in March 1931, Webb bid for a pewter punch bowl, 30 coloured prints, an image known only as *Canadian View* and a set of antiquities.[18] With little detail to go on, it is difficult to track why these objects were acquired or to know if they were part of a more elaborate scheme. What happened to these objects after they left the auction room is unclear.

Webb remarked that 'Wellcome thought up elaborate precautions at auction sales to disguise his interest in articles. Staff used false names and tried to buy many articles which might include only one that was actually wanted. This ploy was soon spotted and resulted in counter-subterfuge to get the price of articles raised.'[19] It is also one reason for the huge amounts of duplication that remain in Wellcome's collection today – for instance there are no fewer than 247 pairs of Spencer Wells-type artery forceps, used to compress arteries, clamp and seal small blood vessels or hold the artery out of the way (fig. 25).

Frank Webb was the Wellcome Historical Medical Museum's longest-serving member of

Fig. 25
Some of the 247 pairs of
Spencer Wells-type artery
forceps in Science Museum
storage.

Fig. 26
Graduation photograph
of Mary Cathcart Borer
in 1928, the same year
she joined the Wellcome
Historical Medical Museum.

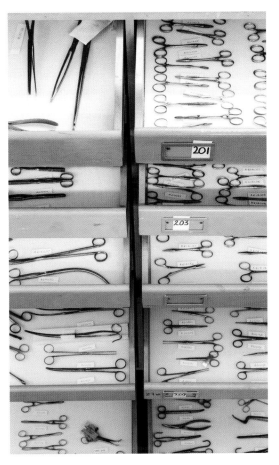

staff, working there for 43 years and ten months, before retiring in August 1965. He had a five-year gap for military service during the Second World War, after which, like many others, he returned to his post at the Wellcome Historical Medical Museum. He also donated two items himself – a castor oil spoon and a pulsometer in February 1950. Both are now in storage with objects of a similar type. Perhaps he wanted to leave a piece of his own medical history in the collection he had served for so long.

Behind the scenes at Wellcome, an army of anthropologists, archaeologists, librarians and cataloguers attempted to bring order to Henry Wellcome's vision. Mary Irene Cathcart Borer (1904–1996; fig. 26) was one such member of staff. After gaining her degree in Geography and Cultural Anthropology from UCL in 1928, Borer answered an advert in *The Times* for 'a graduate in archaeology or anthropology, with knowledge of French or German'.

Like all Wellcome Historical Medical Museum staff, Borer submitted a monthly report of her activities showing the tasks that filled her day (fig. 27). These reports remain in the Wellcome Library archives today and give us an insight into her working life. There are huge similarities with the day-to-day activity that underpins collections work today. Like most of the staff at the Wellcome Historical Medical Museum, Borer's contributions were written silently into displays rather than recognised explicitly. She catalogued whole collections, like those of Winifred Blackman, an agent employed by Wellcome to acquire on his behalf in Egypt.[20] She also wrote and typed labels and carried out research, specialising in prehistoric anthropology and human remains. She even organised a display at the Wellcome Historical Medical Museum for the English Folk Dance Society in 1931.

The last two years of Borer's Wellcome career were spent at Euston Road in the Registration Department. Thousands of items passed through Borer's hands – indeed, most of the objects Wellcome acquired would have been processed by the department. From her surviving ledgers (fig. 28), she appears to have been given the responsibility for metal-worked objects. One such object came via auction having been bought by Frank Webb.

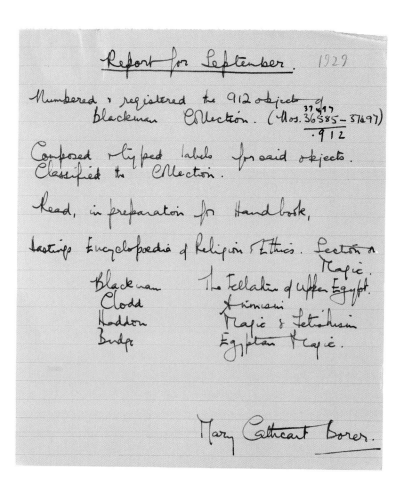

Report for September. 1929

Numbered & registered the 912 objects of Blackman Collection. (Nos. 36585 – 37497)
  37497
 -912

Composed & typed labels for said objects.
Classified the Collection.

Read, in preparation for Handbook,

Hastings Encyclopaedia of Religion & Ethics. Section on Magic.

Blackman   The Fellahin of Upper Egypt.
Clodd   Animism
Haddon   Magic & Fetishism
Budge   Egyptian Magic.

Mary Cathcart Borer.

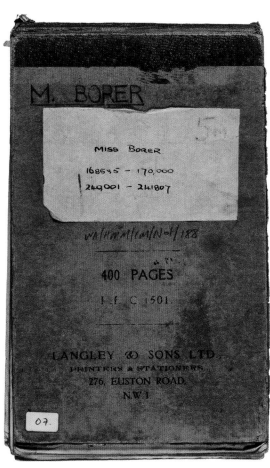

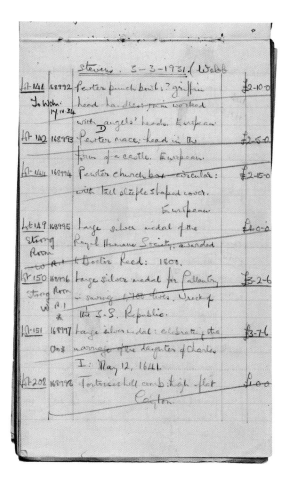

Fig. 27
Mary Cathcart Borer's report for September 1929 detailing her work.

Fig. 28
Borer's notebooks for cataloguing, 1931.

Fig. 29
Silver medal (front and
back views) of the Royal
Humane Society, awarded
to Dr Reed in 1800.

Science Museum Group
Object number A168995

Borer recorded in her notebook the date and place of the auction as 3 March 1931. She gave the object an identifying number so that it could be tracked, described it as 'large silver medal from the Royal Humane Society awarded to Doctor Reed, 1800' and the price paid (£4), before the object was placed into a secure store known as the 'strong room', where it remained until its transfer to the Science Museum. Perhaps time will reveal why Dr Reed was awarded this medal and how it came to be sold (fig. 29).

For working women at the time, marriage bars were frequently enforced in the workplace, including Wellcome's museum and his pharmaceutical business. Once married, women had to resign from their employment. In her 1935 resignation letter to Henry Wellcome, now held at the Wellcome Library, Borer's commitment to the Museum's endeavours is clear, and she resented the marriage bar rule deeply. She wrote saying, 'If during my stay in Egypt this winter, you would like me to collect any material or information for you, I shall be delighted to do so.'[21] Borer went on to work with her husband, Oliver Humphreys Myers, an archaeologist specialising in Egyptology. Myers had donated items to the Wellcome Historical Medical Museum in 1932 on behalf of the Egypt Exploration Society, and it is possible that this is how the couple met. Borer's marriage to Myers lasted only three years. After their divorce, she developed her literary career, writing popular histories and stories for children, under her maiden name. Within the Wellcome Library is a screenplay she wrote about Joseph Lister's journey from junior surgeon to the developer of the antisepsis system in surgery. She was almost certainly inspired by Johnston-Saint's collecting of Lister's personal effects, relics and copies.

At the time of Wellcome's death in 1936, it is estimated that his collection stood at over a million objects – more than that of the Louvre. No one knows with absolute certainty just how many objects Wellcome's enterprises netted. His passion for collecting grew faster than any administrative accounting was able to keep up with. Four storage sites around London housed the burgeoning collection. Not a single person has seen the collection in its entirety – it would have taken a lifetime to even tap the surface.

After he died, Wellcome's body lay in state at his medical museum, guarded over by long-serving members of staff such as Webb. The Wellcome Trust was set up according to the terms of his will, along with a board of Trustees to make decisions about the pharmaceutical business and research laboratories as well as the collection. The pharmaceutical business carried on and the Library material remained at the Wellcome Trust. The Museum collection was seen as something separate – a distinction not drawn by Wellcome himself. The objects that were deemed surplus to its needs or non-medical were put up for sale at auctions or transferred to museums around the world.[22] One of the items transferred was Charles Blondin's wheelbarrow, along with the cord and trapeze, that he used to cross Niagara Falls in 1859. This is now at the Victoria and Albert Museum in London.

Those who had lovingly built up the collection, cared for it and interpreted it for audiences through exhibitions saw much of their lives' work dismantled – so much so that Johnston-Saint told a Wellcome member of staff who was collecting reminiscences that, 'He could not usefully contribute to my interviews because he was so bitter about the way the Wellcome collection was handled during and just after the 1939–45 war. He looked on the collection as quite unique and thought with proper use it could have been the single greatest contribution to the history of medicine.'[23]

After ten years of discussion, the largest transfer of objects was announced in 1976: a long-term loan to the Science Museum. It led to two things: a new curatorial team and new galleries. For the first time, the Science Museum had a dedicated team for medicine. One of its first jobs was to unpack, photograph, catalogue and store the collection – a mammoth task that took four years and over 30 people. Computer records were introduced for the first time, alongside handwritten ledgers to help keep track of the thousands of additional objects now in the Science Museum's care.

The range of material transferred to the Science Museum still surprises people unfamiliar with its history. It includes amulets to protect against aches and pains, surgical instrument sets, stethoscopes, pharmacy jars and prosthetic limbs. Within an institution dedicated to science, it can sometimes feel that religious saints jostle for attention with chemical glassware, and amulets rub uneasily alongside molecular models. Yet while their origins and uses have been different, they share a common thread – all evidence the myriad ways in which humans have attempted to tackle the challenges of well-being and illness.

Two galleries resulted from the long-term loan of the Wellcome Collection to the Science Museum. In 1980 Glimpses of Medical History (fig. 30) opened – a series of room sets and small-scale dioramas that looked at interactions between patients and practitioners. The following year, the Science and Art of Medicine gallery (fig. 31) took visitors on a time-travelling journey from ancient Mesopotamia to the cutting edge of medical technology in the 1980s.

By this time, many new techniques and technologies had developed since Wellcome's death and were conspicuously missing from the collection. Items such as molecular models, heart and lung machines, psychological testing kits, kidney dialysis systems, penicillin, large-scale dentistry kits and many others have now made their way into the Science Museum to try to fill the gaps. As is the case for all major museums, only a very small proportion (around five per cent) of the collection can be displayed at any one time, such is its breadth and depth.

The Science Museum's own collecting of objects and artefacts from medical history increased from this time – an upturn that continues today, though not at the same volume or pace as Wellcome and his agents had maintained. Unlike them, we do not try to be comprehensive because no museum store could ever contain such an encyclopaedic collection – it would simply take up too much space. Where possible, we collect testimony about objects from their users and makers, meaning we can not only tell stories about innovation in medicine but also document people's own thoughts and feelings, uncovering hidden stories. So often, we are separated by time and context from the users of the objects we care for. These personal accounts will hopefully help to bridge that gulf.

Defining the parameters of the Science Museum's collection, setting criteria for acquisitions and having a clear idea of our aims enable us to make decisions about what to acquire. Sometimes we actively seek out objects that are not currently represented in the collection; sometimes we accept objects offered to us by members of the public.

Figs 30 and 31
The Glimpses of Medical
History gallery (above),
which opened in 1980,
and the Science and Art of
Medicine gallery (below),
opened in 1981, both at
the Science Museum.

Fig. 32
Tracey Baynam's prosthetic
legs for use from the age
of eleven months to her
teenage years.
Science Museum Group.
Object numbers 2017-75 to 79

Fig. 33
Tracey wearing her
prosthetics as a child.

The mission of the Museum and its sister museums in the Science Museum Group is to 'consistently provide the nation with the world's best material and visual record of science and technology and its impacts, including industry, medicine, transport and the media'.[24] This is a responsibility held by all of our curators but particularly by those who work with Wellcome's collection, who have inherited a legacy broad in scope and in a subject as emotive as medicine. Medicine touches all our lives – we are deeply concerned with our own health and that of our family and friends. As curators, it is not uncommon for our own medical experiences to make us think about collecting. Any of our own stories could be similar to those of our visitors. When swine flu became headline news in Britain in 2009, one curator brought the Tamiflu medicine prescribed for their son into the collection. Tamiflu had a mixed reception amongst medical professionals and members of the public since there were questions over whether the benefit of the medicine outweighed its unpleasant side effects, which included nausea, vomiting and headaches. Many people who were prescribed Tamiflu decided not to take it or did not finish their course of tablets.

Likewise, the history of medicine covers its negative impacts as well as its successes. In 1999 the Science Museum was offered a large collection of prosthetic limbs, appliances and booklets by Dr Ian Fletcher, who was based at Queen Mary's Hospital in Roehampton. The hospital was looking for a new home for the objects owing to constraints on their existing space. Contained within this set of 258 objects were prosthetics made for children affected by the drug thalidomide.

Thalidomide became available in Britain from 1958 to 1962 to treat a wide range of conditions, from colds and flu to morning sickness in pregnancy. Britain was just one of 46 countries where the drug was prescribed. It was marketed as completely safe for all and could be found in hundreds of products worldwide, under several different names, although some early users reported numbness and pain in their hands and feet.[25] Soon, however, there was an increasing number of reports of children being born with under-developed or missing limbs. The link between the drug and the impact it had on foetal development took years to uncover, in part because the effect occurred only when mothers-to-be took the the drug during a very specific period of their pregnancy (20 to 37 days after conception). Each of these seventeen days impacted a different part of body development from ears, arms or legs to internal organs. It is estimated that thousands of children worldwide were affected. In light of thalidomide's effects, drug regulation was tightened to prevent such an unexpected outcome happening again.

When the Science Museum was asked by the Thalidomide Society if we would like to acquire seven sets of prosthetics used by Tracey Baynam, whose leg development was affected by the drug, it meant that for the first time we could trace one person who was affected by thalidomide's story over several years (figs 32 and 33).

For years, the Thalidomide Society have been recording testimonies of people's experiences of thalidomide and they have kindly shared Tracey's story with the Science Museum.

For Tracey, each pair of leg fittings required four or five visits, followed by an annual visit, to her prosthetic fitter, Fred. She has said of her legs, 'The trouble was they were like an old comfy pair of shoes. By the time you'd got used to them and they were all comfortable you were too big for them and you'd got to have another pair.'[26]

For one set, she persuaded Fred to make her a set of prosthetics to accommodate four-inch stilettos. On another occasion, she wanted to be slightly taller as her new boyfriend was tall. Fred accommodated her requests, sometimes working in his spare time. Many children affected by thalidomide rejected prosthetics, preferring to manage with their own limbs, though Tracey said, '[I] wanted to [wear the limbs] because I was trying to be normal and a normal person wasn't in a wheelchair, a normal person was standing upright'.

After years of walking with sticks to help her balance on her prosthetics, Tracey developed osteoporosis in her shoulders. She had also married and had an active young family and decided then to start using a wheelchair. Occasionally, she still wore her prosthetics while using her wheelchair, to 'appear normal', in her words. After a while though, the prosthetics 'hung around into the bedroom for ages and ages. Then they just got moved to the loft … It wasn't done lightly … I needed to put them out of sight so that I'm not tempted because they hurt my shoulders.'[27]

Capturing how people feel about donating their items is rare and we are fortunate that both Tracey and the Thalidomide Society have shared their feelings about it with us:

I'm very happy and proud that my legs are at the Science Museum because it will be nice for my grandchildren and even great grandchildren to be able to see them long after I'm gone. Hopefully they will be interesting and educational for people visiting the Museum wanting to know more about thalidomide.

Tracey, 2018

Although they have been lying silently in a loft for so many years, they have now suddenly been reborn – firstly by being collected by the Thalidomide Society, then making their way onto the set of TV show *Call the Midwife* and now in their new home, the Science Museum. So many objects like

this are thrown away and I feel so lucky that they have been saved to give a glimpse into the life of Tracey and also the lives of so many other thalidomide survivors across the UK.

Ruth Blue, Secretary of the Thalidomide Society, 2018

Curators also actively seek out items to add to the collection. The bans on smoking in enclosed public spaces and the workplace that were introduced across Britain during 2006/2007 were a watershed moment in public health. The bans were the first time that legislation had been passed to prevent non-smokers being affected by second-hand smoke. According to the National Health Service (NHS), people who regularly breathe in second-hand smoke are more likely to get the same diseases as smokers, including lung cancer and heart disease, than those who are not exposed to it. At the time it was controversial, with opponents to the ban arguing that the risks of passive smoking were overstated and that the civil liberties of smokers were being impinged.

Fig. 34
Local and national awareness campaigns from the smoking ban in England and Wales, 2007.

Science Museum Group.
Object number E2010.39

Shortly before the ban came into force in England on 1 July 2007, Stewart Emmens, the Science Museum's Curator of Community Health, set out to collect the ephemera that would soon disappear from restaurants, cafés and pubs – ashtrays, smoking-policy signs, as well as local and national information and awareness materials produced at the time (fig. 34). Emmens's association with the Science Museum's Medicine collection began with the unpacking of Wellcome's collection and he has spent his career adding to it.

Stewart undertook a letter-writing campaign, contacting the owners of large pub chains to try to elicit various pieces of smoking-related paraphernalia likely to be discarded after the bans. On the eve of the ban in England, he made visits to local pubs, restaurants and other establishments to pick up objects to add to the growing haul (fig. 35). While it is established practice to document the reasons for acquiring objects, I was able to ask my colleague about his motivations directly. Stewart recalls:

This felt like a pivotal moment in public health that needed to be marked. Familiar environments were about to be changed overnight and it seemed that there was a danger that a whole series of objects would be lost as businesses cleared the decks in order to comply with the new regulations. Any existing smoking-policy signage, for instance, effectively became illegal and had to be removed from sight. I wanted to ensure that some of it was preserved. Another aspect of the collecting project was to reflect on the bewildering range of awareness materials being produced in the run up to the bans – from posters and pencils to soft toys and t-shirts. A selection of such objects, produced *en masse* for these one-off events, are now held within our collection.[28]

More than ten years on, the impact of the ban has been millions fewer smokers and fewer cases of bar workers with respiratory problems. Only time will reveal the long-term health effects but public attitudes to smoking now favour further measures to reduce the impact of smoking.

Fig. 35
Glass ashtray from the Groucho Club, Soho, London, said to be the last such ashtray on the premises not removed in the run-up to the smoking ban, 2007.

Science Museum Group.
Object number E2010.39.32

There are thousands of stories waiting to be discovered within the Science Museum's medicine holdings, including those from Wellcome's unique collection. This chapter has barely scraped the surface of the rich resource of objects and the archives that go with them. Knowing the original intentions behind some of their acquisitions reveals hidden stories about objects' journeys that we can also use to inform our speculations about those without documentation.

Knowing the different routes by which objects have come into the Science Museum's Medicine collection can help us to understand some of the objects that do not have such a paper trail. One such favourite object amongst curators at the Science Museum is a netsuke in the form of a rabbit using a pestle and mortar (fig. 36). Measuring just over three centimetres high and carved from ivory, this tiny object probably refers to the Japanese myth of the 'Hare in the Moon', a messenger of the moon deity who mixes the elixir of immortality with his mortar and pestle. Netsuke are toggle-like ornaments, designed to secure hanging objects such as medicine boxes or tobacco pouches from the sash of a kimono. It may be that this one was acquired as much for its representation of the pestle and mortar as for any peripheral medical function it may have performed.

Many of the people mentioned in this chapter and others who have worked with objects either at Wellcome Historical Medical Museum or the Science Museum have career-long associations with them. Curating is a unique opportunity to be part of something remarkable, to add to the record of the history of medicine. Each object has its own rich tapestry woven between the lives of the people it has touched. And when an object enters a museum collection, a new chapter begins. Being one of the actors in its story is what makes being a curator a privilege. Henry Wellcome's business acumen combined with his zeal for collecting gave subsequent generations of curators at the Science Museum exceptional material to work with and build upon. Each of the people who have collected, used and preserved objects has shaped the stories that can be told today. In turn, the decisions made by today's curators will influence the stories that tomorrow's curators will be able to share with visitors about this essence of all our lives – our health.

Fig. 36
Netsuke in the form of a rabbit grinding with a mortar and pestle, ivory, Japan, 1701–1900.

Science Museum Group.
Object number A641096

14509.

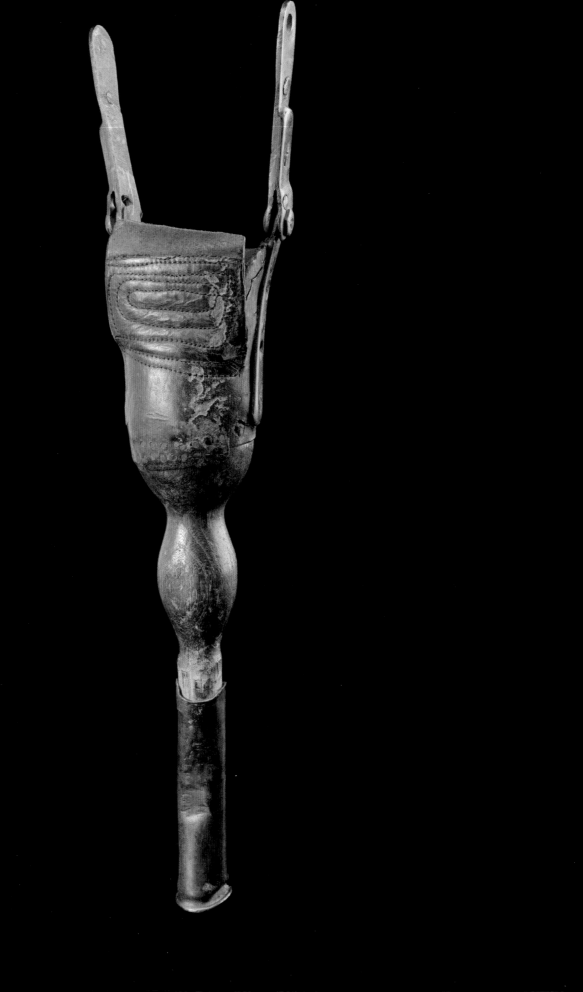

2
PROSTHESES AT THE SCIENCE MUSEUM
PEG-LEGS, PYLONS AND THE PIANIST'S ARM

STEWART EMMENS

Fig. 37
Wooden peg-leg made from a piece of domestic furniture, 1903.

Science Museum Group. Object number 1999-490

Back in 1903, a father fashioned an artificial limb for his three-year-old son (fig. 37). From what was probably a small chair or low table, he took one of the wooden legs and created a tiny limb, to which he added a hinged iron knee joint and a leather knee cap. He also reinforced its lower end with a tubular metal sleeve – a wise decision given the battering the peg-leg seems to have got from the active toddler.

Our records suggest that the father made it at his place of work, a shipyard in the northern English town of Blyth. They also note that many decades later, at the age of 79, his son was still wearing an artificial limb, though one that was rather more conventional than his first. As to whom this father and son were and whether the boy's leg came to be lost through accident or disease, we do not know.

The Science Museum holds one of the largest, and perhaps the best, collections of historic limb prostheses in the world (figs 38 and 39). But as is the case with many museum objects, their former owners are mostly unknown to us. Names and other details may be lost, but as with the young boy's peg-leg, tantalising clues often remain – and these may help us evoke a faint, ghostly image of the limb's now-absent wearer.

For centuries, people all over the world have attempted to replace the lost, missing or defective parts of their bodies. From hip joints to heart valves, breasts to eyeballs, over time the list of options has increased greatly. Depending on what is being imitated, these devices may transform the workings of our bodies, sometimes prolonging

or even saving our lives. In other situations, they can be purely cosmetic – an attempt to restore lost symmetry, boost personal confidence or simply avert the curious gaze.

Artificial limbs are perhaps the most culturally prominent form of prosthesis. As items of material culture, they are both personal and highly intimate, offering tangible reminders of trauma and of absence created by war or accident, disease or genetics. These objects may sometimes reveal wider social issues too, not just the prevailing attitudes towards disability, but also towards age, race and gender. They might highlight a wearer's financial status, since this may determine the degree of access to prostheses as well as the quality of the limbs themselves. In the eyes of the public, those wearing prostheses can also be perceived in various ways, depending on the time, context and circumstances – some may be seen as brave heroes and others as helpless victims.

The general stigmatisation of those who are missing limbs and those who wear artificial replacements also has a long history. Culturally, either the physical absence or the artificial substitute has often been associated with negative characteristics. Be it Robert Louis Stevenson's Long John Silver, J. M. Barrie's Captain Hook or Charles Dickens's Silas Wegg (figs 40–42), the loss of a limb has been presented as a key element that embodies a character's villainy. And these characters were not simply the works of creative imaginations. For example, it appears as though much of the inspiration for Stevenson's

"THIS MAN IS MINE!"

Figs 38 and 39 (opposite)
A selection of the Science
Museum's artificial limb
collection.

Figs 40–42 (above)
From left to right,
Long John Silver, Captain
Hook and Silas Wegg:
notorious amputees
from classic literature.

infamous character came from his friend, the poet and journalist William Ernest Henley. Childhood tuberculosis had led to the amputation of one of Henley's legs and it was a combination of his wooden peg-leg coupled with what Stevenson described as his 'maimed strength and masterfulness'[1] that provided a real-life template for Silver. Whether Henley, a popular, gregarious and jovial man, was truly happy with his repurposing as a murderous villain is unclear.

The majority of the Science Museum's artificial limb collection is in storage, so extensive is its scale. Spanning several centuries, the collection's greatest strengths lie in objects from the late 1800s onwards. It was from the mid-19th century that significant improvements in both the medical and technological management of limb loss were made, in parallel with the gradual emergence of a defined limb-making trade. Increasing numbers of those who lost limbs were surviving. Previously, traumatic limb loss, be it through accident or surgery, had very often proved fatal. The introduction of effective anaesthetics in the 1840s and antiseptic surgical techniques in the 1860s improved a patient's prospects. However, controlling infection

remained a huge problem until the introduction of antibiotics in the mid-20th century. For example, many of the numerous amputations carried out during the First World War were actually pre-emptive surgeries performed to prevent the spread of potentially fatal infectious conditions, such as gas gangrene.

The control of blood loss was also key to a patient's survival chances. Ancient techniques to reduce bleeding, such as cauterising with a hot iron or tying off blood vessels with a ligature thread, endured before transfusion became a viable option in the latter stages of the First World War. The use of tourniquets to stem bleeding is also central to any discussion of both blood and limb loss. It remains a crucial life-saving device today, though in certain circumstances they can be highly dangerous. Many soldiers either died or lost limbs during the First World War after tourniquets, applied in the immediate aftermath of wounding, remained in place for many hours and, as tissues were starved of oxygen, gangrene set in.

The very earliest prosthesis represented in the Science Museum's collections is actually a replica (fig. 43). The original bronze leg dating from around 300BC was excavated from a Roman

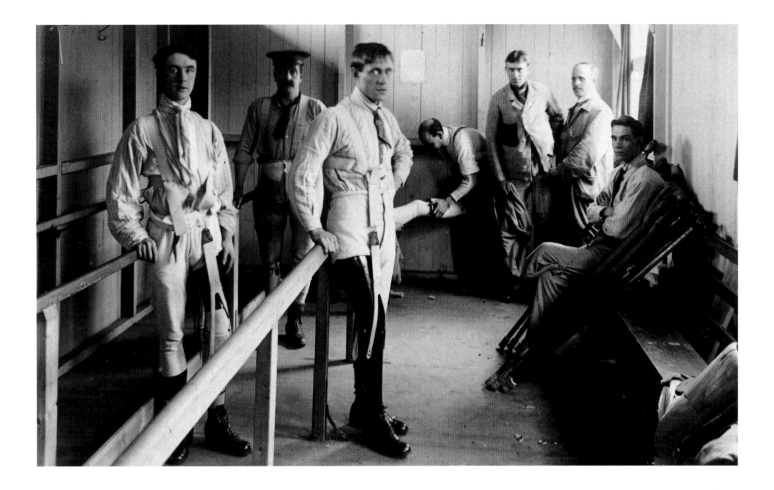

Fig. 43 (opposite)
Copy of a bronze Roman
leg found in Capua, Italy.

Science Museum Group.
Object number A646752

Fig. 44 (above)
Wounded British soldiers
from the First World War,
wearing their new artificial
legs, Queen Mary's Hospital,
Roehampton, 1915–18.
The parallel bars are to
support the men as they
learn to walk.

Science Museum Group.
Object number 1999-317/5

grave in Capua, Italy, in 1858 and was the oldest prosthetic limb to have been discovered. Held by the Hunterian Museum at the Royal College of Surgeons of England, it was sadly destroyed along with numerous other irreplaceable objects during an air raid on London on 11 May 1941. Fortunately for posterity, a copy had been made some years earlier.

War, of course, has always been a major cause of limb loss. This was especially true during the First World War, when over 41,000 of the near two million soldiers who were wounded fighting for the British and Empire forces lost at least one limb. As these amputees returned *en masse* from the battlefields, the existing systems of provision back home in the UK were rapidly overwhelmed. Only with the establishment of a limb manufacturing and fitting centre at Queen Mary's Hospital, Roehampton, in 1915 did this crisis even begin to be addressed properly (fig. 44).

In the late 1990s the Science Museum acquired a large number of historic limb prostheses that had been gathered by Queen Mary's over many years, mostly by Dr Ian Fletcher, a Senior Medical Officer at the hospital. This internationally significant group now lies at the heart of the Museum's medical holdings, an invaluable resource that can illustrate how innovations in technology, engineering and medicine have impacted on limb design and manufacture. It can also reveal poignant insights into the lives and experiences of limb wearers, not only in the First World War era but also in the years before and the many decades since.

These objects physically embody the many challenges encountered when trying to recreate or reintroduce a lost or absent function. The human limb is a highly complex and subtle machine whose workings are not easy to mimic, be they the smooth motion of knee and ankle joints or the intricate movements of hands. However, simply designed prostheses have often proved to be more successful in meeting the needs of the wearer than more complicated varieties. One could conclude that technological sophistication does not necessarily equate to a better artificial limb.

A good example is seen in the case of Carnes artificial arms (figs 45 and 46). Once made in the American Midwest by the Carnes Artificial Limb

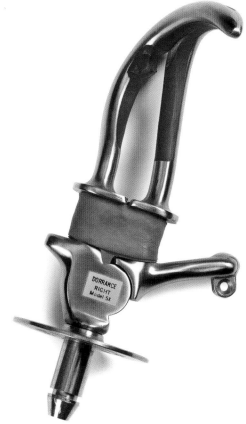

Company, they were highly coveted for a number of years, particularly by First World War amputees. Famous for their imitation of natural movements, the arms offered striking articulation of the metal hand and wrist, and a bending of the elbow. But achieving these actions required tremendous patience and commitment from the wearer. The arms were attached via a shoulder harness, and a whole series of body movements had to be learned rigorously in order to tense and slacken the internal cords and cables that made the arm move.

First produced in 1904, it had established quite a reputation by the time war arrived a decade later. From 1915, British authorities sent patients' measurements out to Kansas City, where arms were manufactured before being shipped back across U-Boat-infested seas to Britain. The Carnes arm was a sophisticated piece of technology, and, as such, an expensive luxury item, outside the typical budget allocated to supply limbs to amputees from the lower military ranks. Only those able to top up the financial allowance with their own private funds tended to be issued with it. As these men were usually officers, it unsurprisingly became known as the 'officer's arm'.

This perceived disparity in provision, influenced apparently by wealth and social class, led to resentment and was even raised in the Houses of Parliament. However, in the end, the Carnes arm was too complex for its own good. It was heavy to wear for long periods, prone to faults and required hours of intensive training (which was not always available during the war years), so many users eventually just gave up in favour of more basic designs or simply reverted to an empty sleeve. Trying to replicate the movements of a human hand has proved an enduring challenge, one that many of the most successful designs have not even attempted. Even today, more than a century on, the basic split-hook attachment (fig. 47) first patented by the American amputee David Dorrance in 1912 remains very popular with many prosthetic-arm wearers. It has always been valued particularly by those returning to work after an amputation because, with this device, they can both grip and manipulate a variety of objects.

Whilst it is undeniable that wars have prompted innovation, developments in limb technology and manufacturing have tended to be enacted quite slowly and unevenly over time.

Figs 48 and 49 (opposite)
Typical wooden leg from
the First World War (above)
and an example of the new
light-metal designs (below)
that were increasingly in
demand by the early 1920s.

Science Museum Group.
Object numbers 1999-444
and 1999-451

Fig. 50 (right)
A well-used wooden leg
once owned by a former
soldier who wore it from
his teenage years until
well into his nineties.

Science Museum Group.
Object number 1999-469

Limb-making was traditionally a craft skill, the creation of a bespoke item for a unique individual. The mass production of limbs, or even a significant standardisation of parts and processes, has proved stubbornly hard to achieve – even when it has been needed most desperately (fig. 48). In Britain, during the First World War, the struggle to keep up with the needs of military amputees reached crisis levels. This was not just an issue of production rates, but also of limb-fitting, medical and social rehabilitation and training veterans how to use their new appliances. Backlogs of patients eventually numbered in the thousands and these would only be cleared in the relative stability of peacetime.

This immediate post-war period was a progressive time for limb-makers. Lucrative government contracts encouraged competition amongst limb-makers, who continued to build on the range of technical advances in limb design and manufacturing that had been made during wartime, despite all the difficulties. It was during this phase that many veterans also began to clamour for the newly emerging light-metal limbs to replace the wooden ones they had first been issued with and that they had perceived as inferior. Such a replacement programme was rolled out across the country during the 1920s and this transitional point is well represented in the Science Museum's collections (fig. 49).

Of course, the many thousands of (mostly young) military amputees often lived on for several decades, during which they could work their way through a number of replacement limbs – not that one prosthesis cannot necessarily serve a lifespan. Amongst the wooden legs we hold is an example that was first issued to a 16-year-old soldier wounded in an earlier British conflict in 1894 (fig. 50). It was then worn daily for nearly eight decades until he very reluctantly accepted a replacement in 1972, at the age of 94.

It was not only the wearer who was of great age: the design he wore is known as the 'Anglesey leg', after Henry Paget, the 1st Marquess of Anglesey. On 18 June 1815, Paget was riding alongside the Duke of Wellington as the Battle of Waterloo entered its final stages when he was wounded severely in his right knee. In what may be an apocryphal exchange, Paget is said to have exclaimed, 'By God, sir, I've lost my leg!' To which, with equal restraint, Wellington replied, 'By God, sir, so you have!' Later that day, Paget's

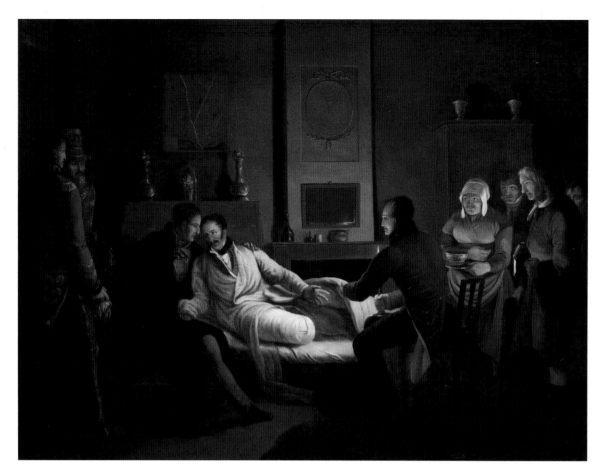

Fig. 51 (left)
Constantin Fidèle Coene,
*Imaginary Meeting of Arthur
Wellesley, Duke of Wellington
and Henry William Paget,*
oil on panel, c. 1820.

Plas Newydd, Anglesey.

Fig. 52 (opposite
and detail below)
A light-metal leg prosthesis
showing extensive
improvised repair work.
The leg was originally
made in Britain around
1934, by the American
company Hanger, Inc.,
which had first set up
workshops here during
the First World War.

Science Museum Group.
Object number 1999-483

limb was indeed amputated (fig. 51). Though he survived, the leg was buried locally and marked by a commemorative stone, and for some years it was a popular tourist attraction.

Our Victorian boy-soldier's leg displays many years of wear and tear, but it remains structurally sound. Even when prosthetic limbs are seriously damaged or seemingly beyond use, owners can often go to extreme lengths to prolong their practical lives. A good fit is crucial for wearers, who can also develop a strong emotional bond with their artificial limb. They can be very reluctant to part with their ageing or disintegrating prosthesis. The evidence suggests that people have sometimes persevered with the help of intense amounts of repair work. A once-pristine metal limb, of the type highly sought after by First World War amputee veterans, is now held together by layers of intricate home repairs (fig. 52).

Why go to such lengths? In this case, we can of course only speculate on the answer, but most probably the owner was afraid that a new limb would be less comfortable than this one. An alternative explanation could be that he was reluctant to risk taking the time off work needed to attend a fitting and then 'wear in' the new leg, which might have led to a loss of earnings or even his job. We do not know, but the leg is a poignant survivor, the physical evidence of one unknown amputee among the many thousands trying to get on with life, but still having to deal with the effects of a wartime wounding.

In the Science Museum's collection, the most extreme example of this 'make do and mend' approach is a leg once worn by another unknown veteran of the Great War. It was issued originally as a temporary limb, designed to be worn for a short period by new amputees as they adjusted physically to their new situation. It seems the wearer was happy enough with this solution and declined to be fitted with the standard prosthesis he was entitled to.

This unjointed cone, made of a fibrous board and known as a 'pylon leg', was subsequently worn for over 40 years (figs 53 and 54). During that time, it was repaired extensively with a wide array of materials including glue, chicken wire and even cement. Originally weighing barely more than a kilogram, it now weighs 14 kg.

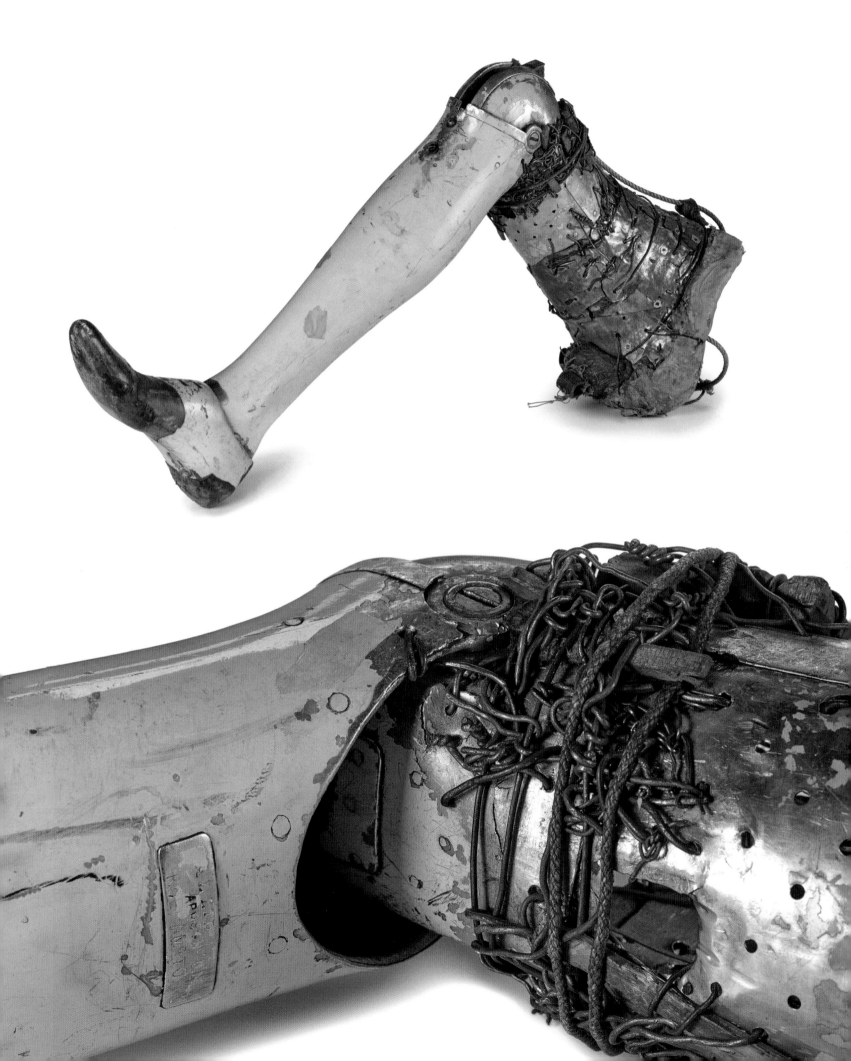

Figs 53 and 54 (opposite)
A heavily repaired leg
(above), worn for many
years by an amputee
working as a thatcher (top).
The temporary cone limb
(below) as it would have
looked originally.

Science Museum Group.
Object numbers 1999-484
and A603150.

So accustomed had the wearer become to this arrangement that when he finally accepted a new limb, it had to be specially fitted with extra weights to reproduce the sensation he had grown so used to. What makes the limb even more remarkable is that its wearer is recorded as being a professional thatcher and roofer, who would have worn this inflexible prosthesis in all manner of adverse weather conditions and on all kinds of uneven surfaces.

As for the roof thatcher, prostheses can offer a welcome way back into employment for amputees. Unfortunately, like others in the disabled community, amputees have often had to campaign and fight for such opportunities. When retraining has been provided for individuals, it has often been into professions very different from those prior to limb loss. At Queen Mary's Hospital, alongside the emergent use of physiotherapy and other forms of physical rehabilitation, a number of former patients from both World Wars were encouraged to became limb-makers themselves (fig. 55). Once trained, they were employed within the hospital's on-site workshops.

In the Science Museum's collection is a set of tools used by just such an ex-military amputee, who was trained for limb-making at Queen Mary's in 1943 and went on to work there for more than four decades. The group of instruments is robust and purely functional (fig. 56). The reinforced-metal arm socket accommodates a series of hammer attachments, with the heaviest weighing over 2 kg. These can be locked into place easily when required for specific parts of the manufacturing process.

However, looking at the design, materials and workmanship of many of the prostheses in the Science Museum's collection, it is clear that specialist limb-makers were not necessarily involved in their manufacture, as evidenced by the young boy's peg-leg. Limb-making as a distinct profession has a short history and for much of this time its products were simply too expensive for many of those who needed them. Traditionally, those seeking a replacement limb would turn to a local blacksmith, woodworker, general handyman or even an armourer. They might even knock something up themselves. This desire to regain lost mobility and

Fig. 55
Former Sergeant Phillips operating a lathe during the production of new artificial limbs, probably at the limb-making workshops at Queen Mary's Hospital, Roehampton, 1944.

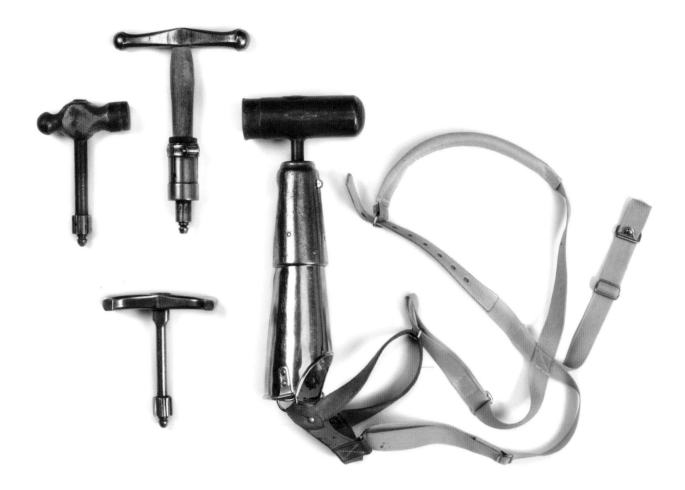

functionality despite limited resources continued and has created some of the most extraordinary prostheses in our collections.

A small and very special group of limbs was made for amputees held in various prisoner-of-war camps during the Second World War. The most striking is for a prisoner with a right-leg amputation above the knee, and its very distinctive painted flesh tone gives a hint to its origins (fig. 58). It was made in the notorious Changi Prison complex in Singapore, where thousands of civilians and Allied soldiers were held following Singapore's occupation by Japanese forces in 1942. The limb was painted to match the remaining leg, whose owner would probably have been wearing shorts or a loincloth. Part-constructed of metal from a crashed aircraft, it was made under the direction of Julian Taylor, a leading surgeon imprisoned in Changi for the duration of the war (fig. 57).[2]

Significantly less sophisticated, but equally illustrative of an amputee's drive to regain independence, is the crudest of improvised limbs, made for a civilian victim of a landmine explosion. After three decades of conflict, during which millions of landmines were scattered across the country, Cambodia has a legacy of tens of thousands of amputees.

This well-worn improvised metal peg-leg, said to be made largely from the cooling system of a vehicle-mounted machine gun (fig. 59), illustrates the lengths one Cambodian amputee went to in order to regain mobility. Fortunately, it was eventually exchanged by its user for a limb similar to that shown adjacently (fig. 60). The simple, hard-wearing and low-cost prosthesis was made by local workers in Cambodia under one of several charitable initiatives providing for what sadly remains a steady annual flow of new amputees. Such local solutions owe much to the earlier development of a form of prostheses that became known as the 'Jaipur limb' (fig. 61). First made in the late 1960s in the Indian city that gives it its name, it was promoted as an alternative to the artificial limbs being imported from America and Europe at that time. Western prostheses were unsuited to the needs of the Indian population, given that many locals went barefoot, tended to squat or sit cross-legged and worked on uneven agricultural land, which was

Fig. 56
Set of tools and re-enforced arm socket used by an amputee employed in the limb-making workshops at Queen Mary's Hospital, Roehampton, c. 1943.

Science Museum Group. Object number 1999-611

Fig. 57
Murray Griffin, 'Artificial limb factory' at Changi Prison, Singapore, ink and wash over pencil on paper, 1943.

OVERLEAF
Figs 59 and 60
A crude 'homemade' metal peg-leg (left) made from salvaged materials and worn by an amputee until replaced by a newer limb (right), as supplied by the Cambodia Trust in 1998, one of a number of similar charities working in the region.

Science Museum Group. Object numbers 2004-39 and 2004-37

Fig. 58
A leg made of scrap metal for an amputee inmate of Changi Prison, Singapore, c. 1943.

Science Museum Group. Object number 1999-429

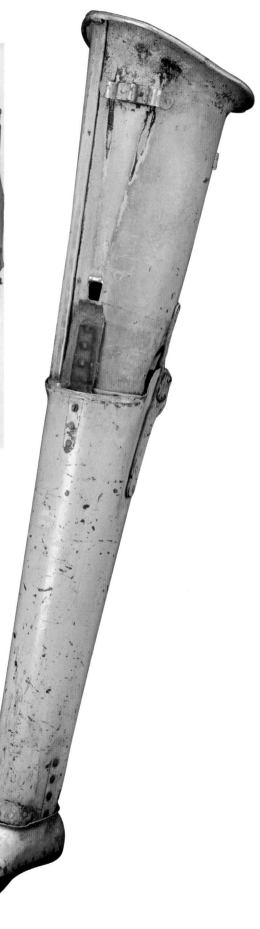

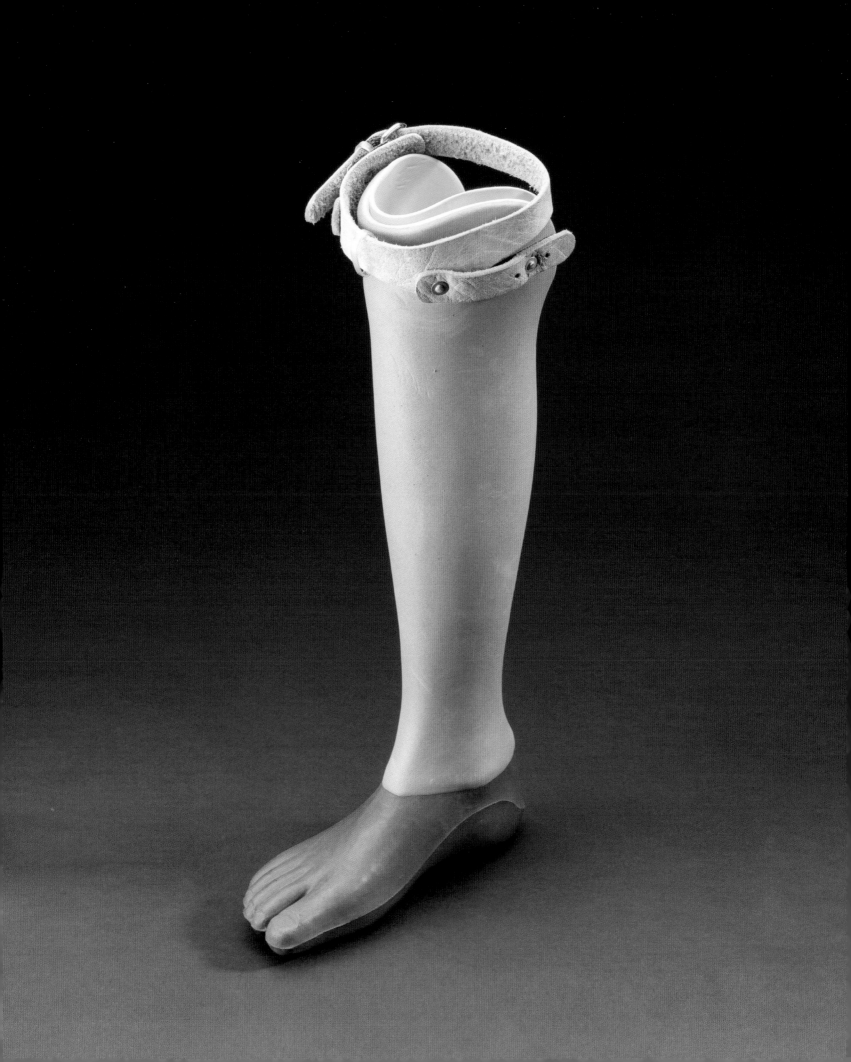

often waterlogged. The expensive imports were just not up to the job.

Designed by a small team of Indian doctors and constructed of rubber, plastic and wood, the Jaipur limb has proved to be enduringly popular. Sturdy, waterproof, better suited to the local terrain and with the flexibility needed for an active lifestyle, the limbs were also given a more appropriate skin tone than the pale foreign imports. Since their initial introduction, well over a million have been distributed worldwide.

The Science Museum's collections contain a disproportionate number of artificial limbs that are associated with military conflict, a fact reflected in this narrative. Given Queen Mary's Hospital's origins and crucial role in the two World Wars and beyond, this is perhaps unsurprising. Many of the limbs in the collection were deemed worn out or obsolete and were simply left at the hospital by patients after they had a new limb fitted. A large proportion of these patients are recorded as being military veterans and we know that nearly all of them were men. Nevertheless, and despite the anonymity of most of the former owners of our limbs, there are examples in the collection that we know definitively to have been worn by women. One of these limbs is perhaps the most unusual of all.

Catalogued as an 'arm made in 1903 for a lady pianist', it has two outer digits that are rigidly spread to cover precisely an octave on a piano's keyboard, while the inner three digits are foreshortened to avoid touching the piano's keys (fig. 62). The two playing fingers were given additional padding that shows evidence of significant use. Our records suggest that this very niche prosthesis was worn for a concert performance at London's Royal Albert Hall in 1906. However, it was only after its appearance on national television that the identity of its former owner was revealed.

Elizabeth Burton (née Wright) was born in Daventry, Northamptonshire in the early 1860s. She became a teacher of music and singing and advertised her services in her local area (fig. 63). At what point in her life she lost her arm remains unknown, but commissioning such a specialised device certainly suggests that she carried on teaching afterwards. With six children to support and the early death of her husband,

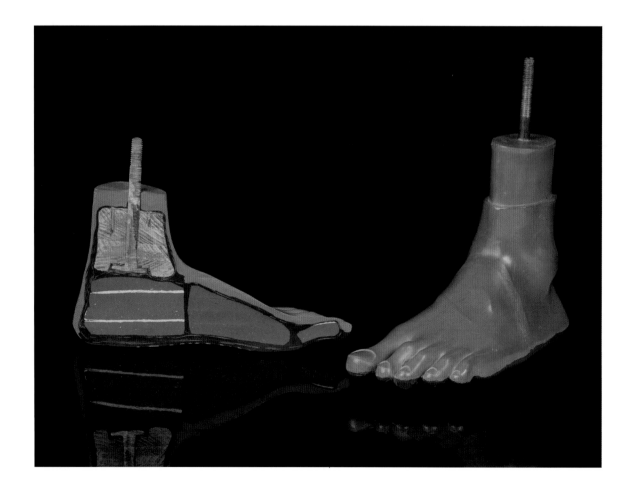

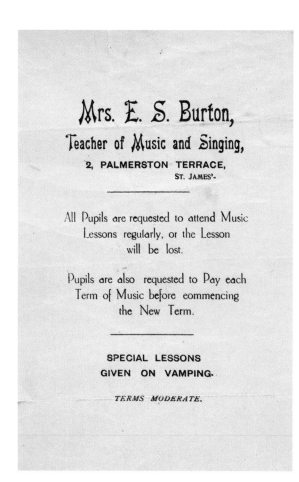

Mrs. E. S. Burton,
Teacher of Music and Singing,
2, PALMERSTON TERRACE,
ST. JAMES'-

All Pupils are requested to attend Music
Lessons regularly, or the Lesson
will be lost.

Pupils are also requested to Pay each
Term of Music before commencing
the New Term.

SPECIAL LESSONS
GIVEN ON VAMPING.

TERMS MODERATE.

Fig. 63
Elizabeth Burton advertised her music teaching on this paper flyer. This included special lessons on 'vamping', a technique that involves repeating a single chord or a simple progression for an extended amount of time. This may well have suited the inevitable limitations of her artificial hand.

Fig. 64
Elizabeth Burton and her daughter Ada, studio portrait, c. 1920.

while she was still in her thirties, continuing with her music lessons was probably a financial necessity.

The story of how Elizabeth came to lose her arm says much about the medical advances we enjoy today, as well as the random nature of events and accidents leading to limb loss. Whilst preparing raw fish for a meal, Elizabeth ran a fishbone deep beneath her thumb nail. An infection set in and, in those pre-antibiotic days, her right hand and much of her forearm had to be amputated in order to save her life.

Pictured later in life with her youngest daughter, Ada (fig. 64), Elizabeth kept her right arm hidden. Within the arrangement of this posed studio portrait the absence is barely noticeable but, as a woman, she may well have felt an extra pressure to conceal her prosthetic limb. Elizabeth had grown up in an era when marriage and motherhood were social norms women were expected to aspire to and then fulfil. The perceived physical imperfection of an amputated arm would have been viewed as potentially disastrous. In addition, the long-discredited belief that various disabilities,

including limb loss, could be hereditary (transferred to offspring via so-called 'maternal impression') remained in popular thought throughout the 19th century, and this could make the situation for female amputees even worse. Although Elizabeth married, had children and pursued her own career, she may still have felt the need to disguise a disability viewed as less acceptable for women than for men. Limb loss has carried an enduring stigma that has been very slow to recede.

A series of posed studio photographs of limb wearers from around 1900, used by the manufacturer James Gillingham for promotional purposes, are certainly suggestive of these differences in attitudes between the genders (fig. 65). While the men appear happy to show their faces in the photographs, many of his female customers have kept a much lower profile. As another leading manufacturer of artificial limbs in this period, George E. Marks, commented in 1888, 'it is very well understood that young ladies wearing artificial limbs are not over-desirous of having it publicly known'.[3] But while disguise and concealment were then

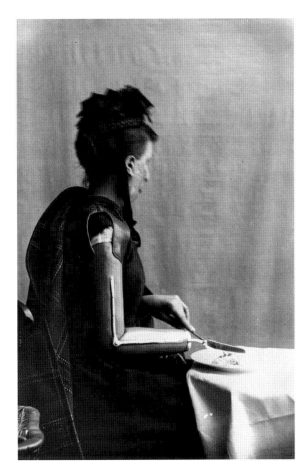

Fig. 65
Clients of the limb-maker
James Gillingham, studio
portraits, 1890s–1900s.
This selection has been
reproduced from glass-
plate negatives in the
Museum's collection.

Science Musuem Group.
Object numbers 1979-191
/21, /27, /39 and /40

seen as particularly important for women, functionality of a limb for domestic work was certainly desirable too.

Artificial-limb wearers have often had little choice about whether to disguise or conceal their prostheses, since within a working environment, practical constraints may have prevented it. And yet a worker who may have spent the day bolting a series of tools into an arm socket may well have attached a non-functioning, more lifelike artificial hand before heading to the pub to socialise afterwards. The desire to restore symmetry with a remaining limb or simply draw less attention to a prosthesis is still in evidence in a number of the more personalised limbs in the Science Museum collection (figs 66 and 67).

With the increasing application of new synthetic materials after the Second World War came the potential to mimic more accurately the look and, eventually, even the feel of a natural limb. This included the ability to match different skin colours and tones, an important feature that could give the limb wearer greater confidence, if only through reducing the disparities between the natural and the false (fig. 68).

During the 21st century, prosthetic technology has made major advances, particularly in the applications of computer technology, neuroscience, robotics and materials science. At the cutting edge is an emerging generation of devices that allow patients to use their own thoughts to create unprecedented levels of movement control and

Figs 66 and 67
This wooden hand (above) from c. 1900 remains encased in its tight-fitting leather glove, while the metal leg (below) from the 1940s retains a laced-up boot and black stocking.

Science Museum Group. Object numbers A653506 and A500467

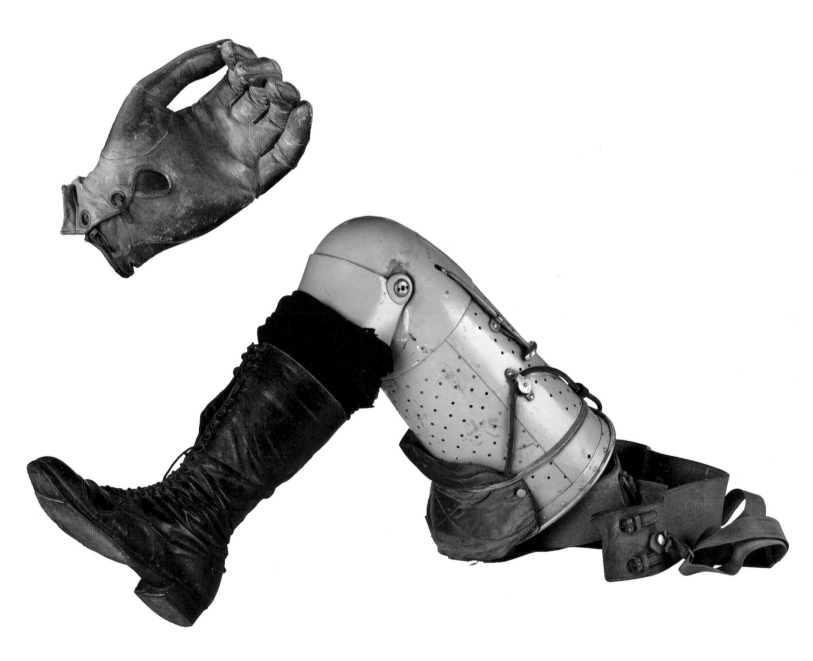

Fig. 68
A plastic arm, made for
a young child missing its
left forearm from birth,
was an early attempt to
match a wearer's skin tone.
RSL Steeper Ltd, c. 1970.

Science Museum Group.
Object number 1999-563

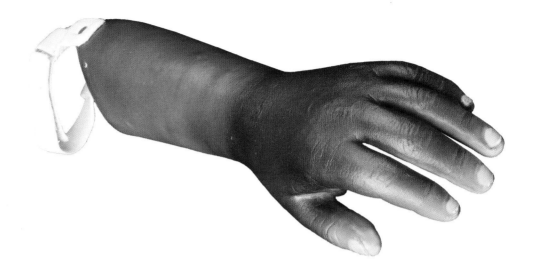

touch sensation. Such technology is also addressing some of the psychological and physiological challenges of limb loss. For example, most amputees endure some level of so-called 'phantom limb pain', a condition in which sensations are felt at the site of the missing limb. This has long proved notoriously difficult to address, but limbs have been developed recently that enable direct sensory feedback between themselves and the wearer's brain. Using what is known as 'myoelectric technology', these prostheses draw on the same neural impulses that once controlled the natural limb and have been shown to reduce the sensation associated with the phantom limb.

The use of high-definition cosmetic silicone also means that limbs can now be extremely lifelike, with synthetic skins for artificial arms and legs, complete with hairs, freckles and other blemishes. Conversely, technical innovations are also offering wearers the opportunity to deliberately draw attention to themselves. A potentially limitless range of bespoke limbs, often highly unconventional in appearance, are now being made to look radically different from both the standard designs and the human limbs they are replacing. This has only a niche market but remains important, since the wearers of these limbs see their prostheses as an opportunity to express their personalities (fig. 69). As an athlete, actress, model and double amputee, Aimee Mullins has stated 'a prosthetic limb doesn't represent the need to replace loss anymore. It can stand as a symbol that the wearer has

the power to create whatever it is they want to create in that space'.[4]

We have also seen the increased media exposure of limb wearers, be they Paralympians, recent military veterans or other individuals, featured on television or online. This could be seen as a part of the drive for greater inclusion across certain sectors of society, but these generally positive representations do not necessarily mirror the experiences of many of those wearing prostheses. As in the wider disabled community, welfare funding and other government policies can have major impacts on the practicalities of daily life, while some prejudiced attitudes within society remain entrenched and difficult to shift. Access to expensive 'state of the art' prostheses is also limited and the technology itself is often unsuited to the needs of the wearer. Across much of the world, limb loss caused by trauma or cancer is now in the minority, and most amputations are due to vascular diseases and associated problems such as diabetes. Amputations as a result of these are often on elderly patients or those with limited mobility. In many such cases, no form of artificial limb may be appropriate, let alone the most advanced designs – which are unaffordable for many wearers anyway.

Two more recent acquisitions for the Science Museum collection reflect both the complexities and ambiguities associated with artificial-limb wearing – and collecting – in the 21st century.

The first, the BeBionic hand (fig. 70), is an example of a myoelectric limb at the pinnacle

Fig. 69 (three views)
Described as a 'retro futuristic leg', this is one of several limbs produced as part of the Alternative Limb Project, founded by the artist Sophie de Oliveira Barata, 2014.

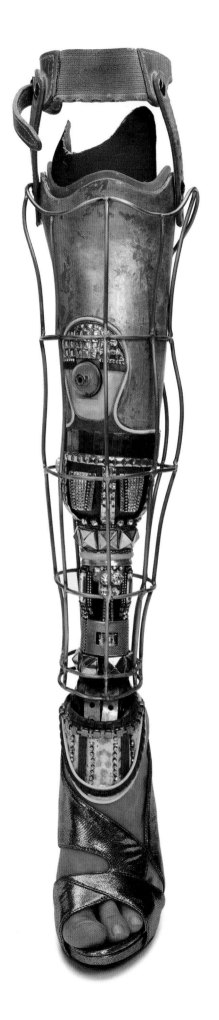
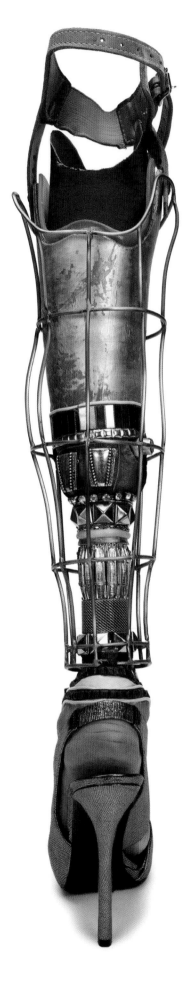

of an evolving prosthetics technology. By means of sensors that detect muscle activity in the remaining arm stump, the wearer can learn to perform a range of movements including complex and very subtle manoeuvres. In some ways echoing the marketing of the Carnes arm a century ago, the BeBionic's makers claim that it can manage many of the everyday tasks one might expect from a biological hand, from facilitating eating and carrying, to switching on lights and typing on a keyboard. Users also have the option of a synthetic covering to produce a more lifelike appearance. With a century of research and development between them, it can at least be expected that wearers of this new technology will be less frustrated by it and therefore less inclined to give up and discard it as many of their predecessors did with those earlier prostheses.

The other new addition to the collection is an example of probably the most high-profile prosthetic limb of recent times – the running blade. Although it has a foot, socket and knee joint, it makes no attempt to imitate the appearance of a human leg. This device is purely for function and performance. Since the design concept was first realised by the American amputee and inventor Van Phillips in the mid-1980s, variations have proliferated and been central to the growth and increasing popularity of disabled sports. This example (fig. 71), launched days before the London 2012 Paralympics, encapsulated much of the advanced technology showcased at the games, but within a more affordable budget.

Though these two new limbs are important additions to the Science Museum's collections, being brand new, they lack the patina of use and the sense of personal ownership that imbue many of the artificial limbs we have. The two objects also represent the more exclusive and expensive end of artificial-limb provision. But while initially they might appear to be light years on from the likes of the thatcher's leg and the toddler's limb made from a piece of household furniture, they too reflect on many of the themes inherent across the Science Museum's wider collection of prostheses. There is the familiar desire of limb wearers for greater independence and the confidence gained through the ability to fulfil

Fig. 70
The BeBionic hand allows the user to learn a range of movements, from the basic to the more complex and delicate. RSL Steeper Ltd, UK.

Fig. 71
Running blade incorporating a hydraulically controlled knee joint and a flexible carbon-fibre foot, Ottobock Healthcare, UK, 2012.

Science Museum Group. Object number 2015-20

everyday tasks without the assistance of others. In ditching the aesthetics of the familiar limb shape in order to compete at the highest levels, today's elite disabled athletes are perhaps echoing the pragmatism of those earlier amputees who often had to fight and persevere to return to the workforce, be they tooled-up craftsmen and machine operators or even piano teachers. It is unavoidable that discrimination, stigma and intolerance are human qualities that are part of our collection's story too. It is to be hoped that as the science and technology of prostheses continues to advance, like the associated management of medical trauma, so society's attitudes towards those wearing them will be equally progressive.

New generations of prostheses look set to emerge over the coming years, which, at the cutting edge at least, will involve an ever-closer merge between the device and its human wearer. We can only speculate as to how the Science Museum's two recent additions will be viewed and valued by future curators – as key milestones along the way? Or as archaic, perhaps barely remembered technologies? Either way, despite their current status, for the foreseeable future most limb wearers around the world will continue to use prostheses that are far less sophisticated and considerably cheaper than the BeBionic hand and the running blade, made to designs that primarily hark back to the simpler, traditional forms of the past. We need to ensure that examples of these most personal of objects continue to be well represented within our Museum collections.

3
THE PROBLEMATIC BODY
DISSECTION AND THE RISE OF THE ANATOMY SCHOOL

ANNA MAERKER

Fig. 72
Wooden reliquary chest,
elaborately carved and
inlaid, containing relics
of several saints, Spain,
18th century.

Science Museum Group.
Object number A634987

How do we know what happens inside the living body? Before the advent of body-imaging technologies such as X-ray and MRI scanning, medical practitioners had to resort to other methods such as feeling the pulse and observing what came out of the body (for instance, the patient's urine). A key method for understanding the bodies of the living was to open up the bodies of the dead, including animals. As we might well imagine, this practice posed many methodological, ethical and legal problems.

Scholars questioned whether one could really learn anything of value about human anatomy from the bodies of pigs, dogs and monkeys, and how much a dead body could reveal about the living. But bodies were always more than just another natural object for scientific study: bodies were associated with personhood. While the practice was contentious due to popular concerns about the resurrection of the dead and the dignity of the deceased, dissection was rarely banned formally in western Europe. Since the Middle Ages, autopsies have been carried out for legal reasons to determine cause of death, and dissections were performed to illustrate and celebrate the body as the pinnacle of God's creation. The bodies and body parts of saints and sovereigns were preserved, displayed and revered (fig. 72).

However, the most visible dissections were performed on the lowest of the low. In the early modern period (roughly from the late 1400s to 1800), doctors were given access to the bodies of executed criminals for public dissections as

a marker of professional privilege for medicine. Such bodies were highly conflicted: on the one hand, they required respectful engagement, while on the other, they were seen as sources of physical and moral pollution. Due to the problematic status of corpses, these public performances entailed the development of sacred rituals to make the opening of criminal bodies respectable.[1] They were witnessed by up to several hundred viewers, including nobility, local dignitaries and clergy. A 'prosector' was tasked with cutting the body open, and individual organs were passed around the audience on trays. Meanwhile the professor stood on a raised lectern, far removed from the messy work of the dissection, and extolled the marvels of divine ingenuity, using the body on display to confirm the knowledge of ancient Greek and Roman medical authorities like Galen and Hippocrates (fig. 73).

Increasingly, however, dissection became more than a mere demonstration of divine creation and ancient wisdom. Andreas Vesalius (1514–1564), a professor of medicine at Padua University, argued that it was crucial for physicians to engage in dissections as a means to produce new insights into the body. In his influential anatomical work *De humani corporis fabrica* (*On the Fabric of the Human Body*, 1543), Vesalius used sophisticated illustrations to highlight the virtues of observation and exhorted his fellow medics to get their hands dirty by performing dissections themselves (fig. 74).

In the centuries following Vesalius's work, physicians increasingly adopted this position and

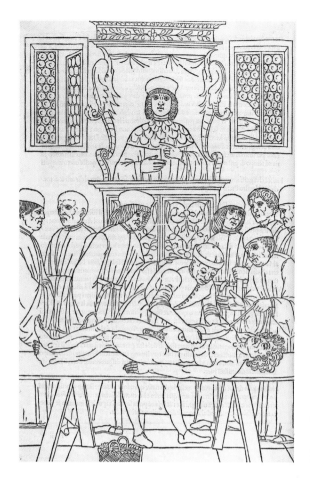

dissection became a central element of medical training for doctors and surgeons. In England, dissections were undertaken by the College of Physicians and the Company of Barber-Surgeons. The dissolution of the Company in 1745 created an opportunity for private anatomy teachers to offer dissections as well. While those lectures were open to any paying customer, they were usually specifically aimed at medical students keen to obtain more experience in hands-on work on dead bodies (figs 75 and 76).

Ritual performances of public dissections were discontinued by the late 18th century as they increasingly came to be seen as vulgar and unscientific. While anatomy teachers remained discreet when it came to the provenance of their material, the practice of dissection itself was acknowledged openly. Dissections were carried out not only in anatomy schools, but also in hospitals and in the homes of deceased patients. Learned journals frequently published accounts of dissections and autopsies of interesting cases and well-known individuals. John Hunter (1728–1793; fig. 77), one of the most eminent anatomists of the late 18th century, testified that

he had himself dissected 'some thousands' of human and animal bodies.[2] He opened a school in Leicester Square in the centre of London that included a museum where specimens of humans and animals were preserved (it was also his home; fig. 78). Such collections of organs and body parts, dried or preserved in spirits, served multiple purposes: they provided teaching materials to illustrate particular pathological conditions, but they also served as collections of trophies illustrating anatomists' social connections to eminent scholars and patients, as well as demonstrating the breadth of their professional experience.

By the 19th century, museums had become central to medical education. Pathological collections were established in medical schools, hospitals and asylums, and served as important markers of institutional and professional identity. As the Edinburgh anatomist Frederick Knox put it, 'Without museums the profession would be in the state of man without a language'.[3]

This increasing demand for corpses, however, far exceeded the number of executed criminals, and teachers had to resort to other, legally and

ethically problematic ways to supply their students with the necessary opportunities to learn. The creation of private anatomy schools in cities such as London and Edinburgh gave rise to the practice of body-snatching: stealing the bodies of the recently deceased from cemeteries and morgues. Such practice, the anatomy teacher William Hunter (John's brother) suggested, was possible practically as long as it was done discreetly so as to prevent public outrage. He said, 'In a country where liberty disposes the people to licentiousness and outrage and where anatomists are not legally supplied with dead bodies, particular care should be taken to avoid giving offence to the populace.'[4]

Dissection was considered offensive as, in popular belief, the physical integrity of the dead body was important (fig. 79). Funerary practices, such as laying out (washing and clothing the body), sitting with the deceased and eating and drinking in their presence, suggest that communities assumed a continuing attachment of the soul to the body. However, dead bodies did not count as property in legal terms. Therefore their appropriation was not considered to be theft, and incurred only a fine, rather than any criminal repercussions. This legal loophole made bodysnatching a lucrative business to gangs of enterprising 'resurrectionists'. Families who could afford to provide protection for their deceased relatives did so, paying for guards to stand watch over the grave until decomposition had rendered the body useless to anatomists, or employing technologies such as 'mortsafes', heavy metal cages to protect coffins (figs 80 and 81). However, unscrupulous anatomists often found ways to obtain desirable bodies regardless. Charles Byrne, an Irishman whose prodigious height of over eight feet earned him the nickname the 'Irish Giant', left instructions to be buried at sea as he was well aware that anatomists were keen to dissect his extraordinary body. However, John Hunter bribed Byrne's undertaker and the body disappeared into Hunter's collection before it could be safely buried.

How could anatomy teachers get around supply problems without resorting to bribery or bodysnatching? Some explored the use of models as substitutes. Midwives, for instance, trained frequently with so-called 'obstetric phantoms', dummies in robust materials such as leather and wood, which allowed practitioners to train in procedures such as turning a breech foetus in the womb (see figs 82 and 83).

Fig. 77 (opposite above)
John Jackson, after
Sir Joshua Reynolds,
John Hunter, oil on canvas,
1813. The skeleton in the
background is believed to
be that of Charles Byrne,
the 'Irish Giant'.

National Portrait Gallery

Fig. 78 (opposite below)
E. Radclyffe after T. H.
Shepherd, *The Hunterian
Museum, Royal College of
Surgeons*, engraving from
*London Interiors with their
Costumes and Ceremonies*,
published by Joseph Mead,
London, c. 1842.

Wellcome Collection

Fig. 79 (above)
Thomas Rowlandson,
*The Resurrection or
an Internal View of the
Museum ...*, pen and ink,
1782. Hunter surrounded by
the bodies of the dissected
on the day of Resurrection.

Wellcome Collection

Fig. 80 (left)
A metal mortsafe used
to protect a coffin from
bodysnatchers, Towie
Churchyard, Aberdeenshire.

Fig. 81 (below)
Mortsafe base and lid,
iron, 1801–22.

Science Museum Group.
Object number A600162

However, such alternatives raised new anxieties. In particular, critics worried that learning with artificial models could lead medical practitioners to lose empathy with their patients and to become accustomed to coarse, insensitive handling.

Models made from alternative materials were problematic for other reasons. Sculptors in 18th-century Europe produced remarkably detailed, accurate and lifelike representations of human anatomy in wax (figs 84–6; see also Chapter 9). However, such objects were closely associated with public spectacle in the form of the popular wax cabinets of the time. Rackstrow's Museum, which was open to the public in the late 18th century in London, for instance, combined wax figures of historical and well-known people with anatomical models and preparations. It vied for respectability after the death of its founder, the artisan and modeller Benjamin Rackstrow (d. 1782), when the museum was taken over by midwife Catherine Clark, who offered medical instruction to accompany the displays. However, throughout the 18th and 19th centuries, such exhibitions were often accused of obscenity and lack of educational purpose.

While audiences were fascinated with such displays, they remained concerned about the appropriation of bodies for medical education and the idea of medical students rummaging among the organs of their nearest and dearest. Public outcry over the scandalous practice of bodysnatching continued and, to make matters worse, there were several cases of actual murder for the purpose of obtaining bodies. The murders committed by William Burke (fig. 87) and William Hare in Edinburgh in 1828, and the killing and sale of the 'Italian Boy' in London in 1831 were highly publicised and caused public panic. As one observer commented, 'The Burkophobia seems to be at its height in the metropolis at the present time; and scarcely a day passes but reports are circulated of the supposed sacrifice of fresh victims to the "interest of science".'[5]

In an attempt to reduce the market for stolen bodies, a government select committee recommended that Britain adopt the French model, making the unclaimed corpses of deceased patients from public hospitals and workhouses available to medical schools for teaching and research. The government passed the Anatomy Act in 1832.

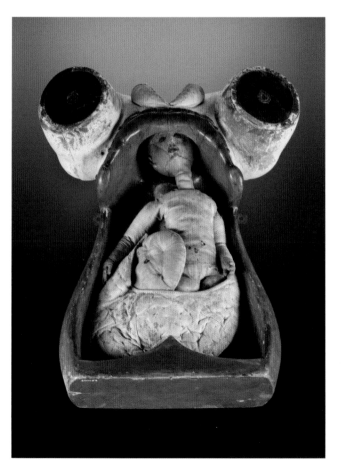

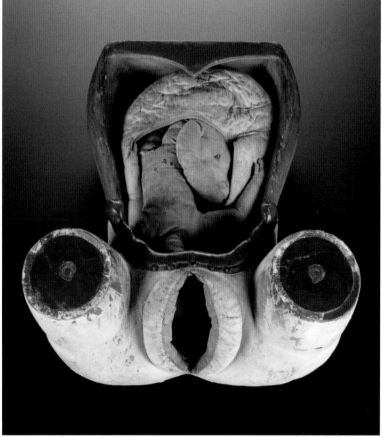

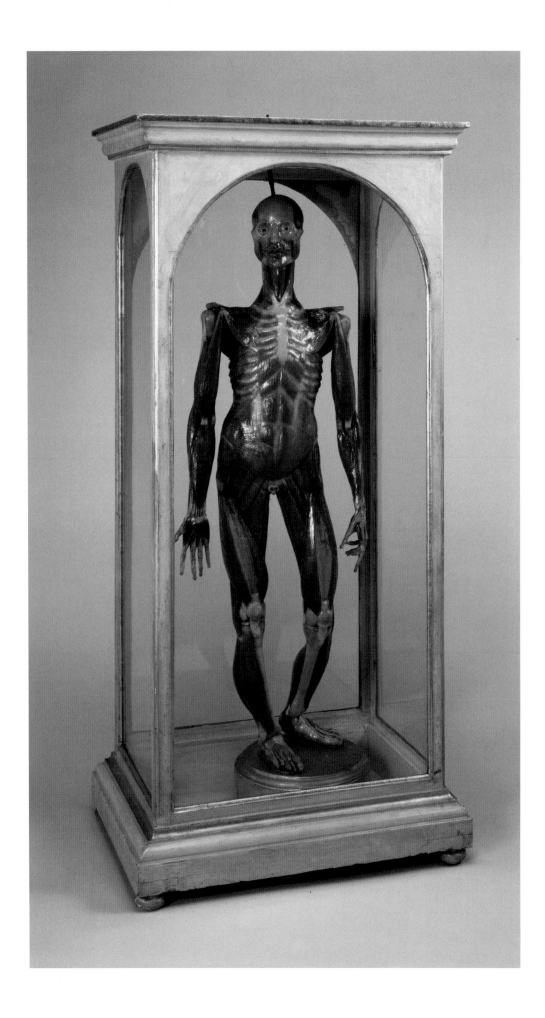

Fig. 84
Possibly Clemente Susini,
Anatomical male figure,
wax, Florence, Italy,
1776–80. Models of the
human body were popular
as teaching tools and
public entertainment
in 18th-century Europe.
Science Museum Group.
Object number A608367

Fig. 85
Anna Morandi Manzolini,
Anatomical male figure,
wax, Bologna, Italy,
1740–80.
Science Museum Group.
Object number A600129

OVERLEAF
Fig. 86
Francesco Calenzuoli,
Anatomical female figure,
Florence, Italy, 1818. The
model shows the internal
organs, with an entirely
removable heart.
Science Museum Group.
Object number 1988-249

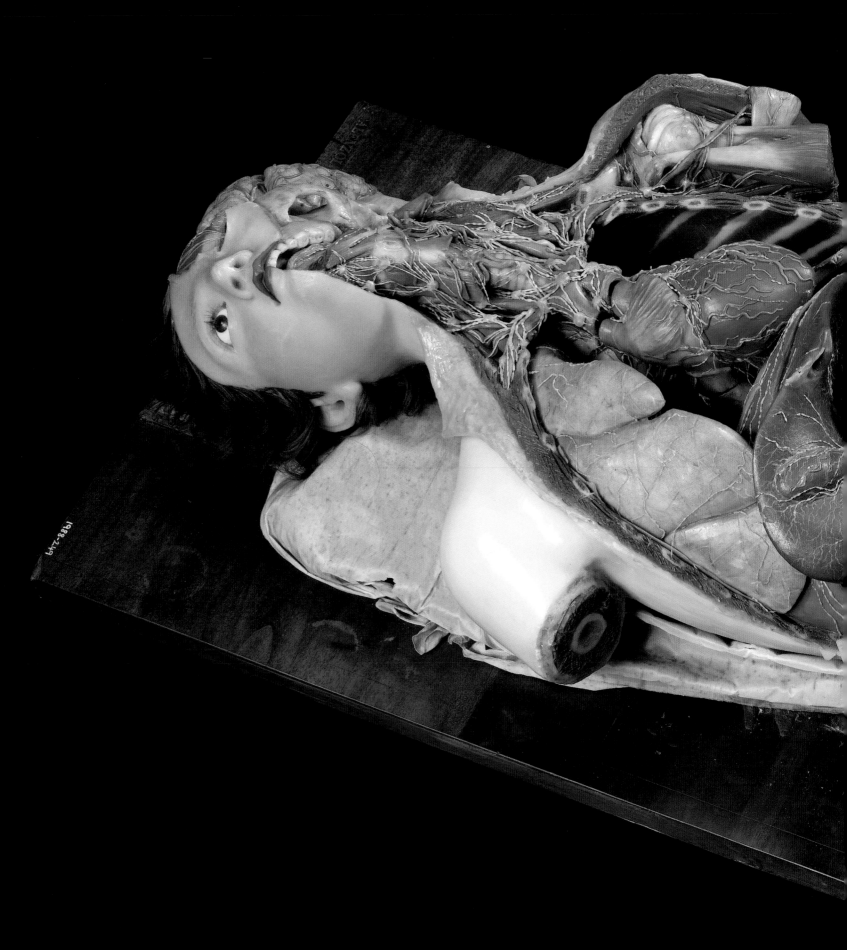

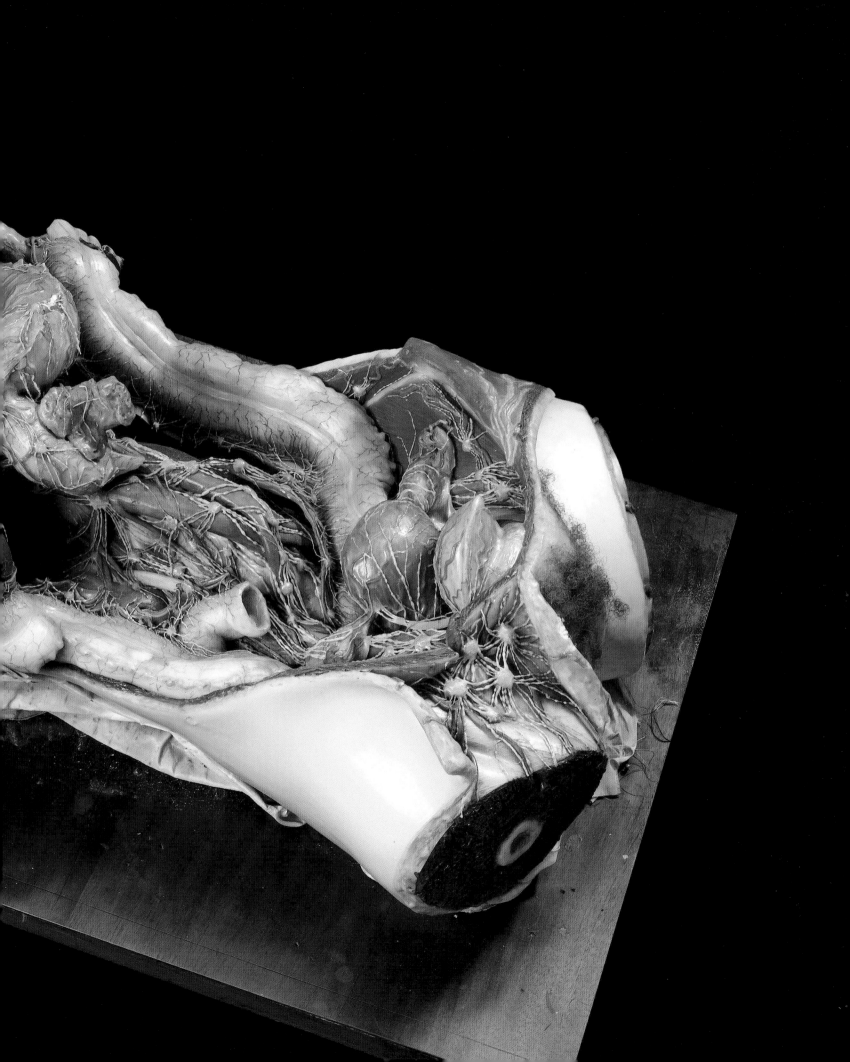

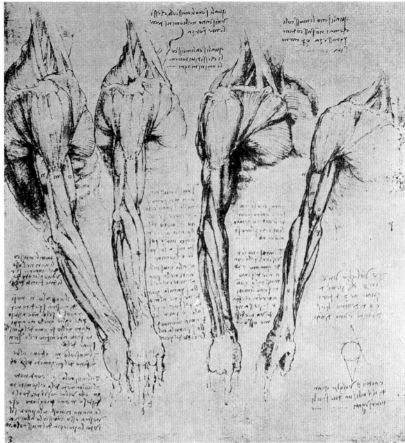

However, since dissection had long been associated with capital punishment for the very worst crimes, there was considerable discomfort with the idea that 'What had for generations been a feared and hated punishment for murder became one for poverty'.[6]

Against public resistance to the idea of making the bodies of dead paupers available for medical training, teachers argued that this practice was a necessity. If students could not practise on the dead, they would carry out examinations on the living patient, often with horrible results. The Scottish surgeon Robert Liston warned that

Many poor creatures have been sacrificed in consequence of the ignorance, carelessness, and self-sufficiency even of scientific professors, who have either despised or neglected the study of surgical anatomy, the consideration of the casualties which may arise during the various operations, and the due education of their fingers. The infliction of unnecessary pain, through want of adroitness in the use of instruments ... cannot by any means be palliated or defended.[7]

Still, supplies remained low and bodysnatching continued in the 19th century, even in France and Britain where unclaimed bodies were legally available. Gangs often raided the burial grounds of underprivileged communities: the poor were targeted, as were the cemeteries of African-Americans in the USA.

In addition to the political, legal and ethical issues, dissection was problematic on practical grounds as well. An ongoing issue for medical students and researchers was the physical disgust they often experienced. The Renaissance artist and polymath Leonardo da Vinci considered his work with 'corpses ... horrible to behold' (fig. 88).[8] To alleviate this repulsion, dissections began with the body parts that decay fastest, such as the intestines, and students attempted to mask smells with vinegar, perfumes or incense (and later with cigarettes and cigars). Medics framed the practice of dissection in a language of heroism. The French anatomist Vicq d'Azyr, for instance, described it as 'painful and perilous': a heroic deed in the service of improving human knowledge and alleviating human suffering.[9]

Fig. 87
The murdering body-snatcher William Burke was himself dissected after his execution in 1829. This small sealed specimen jar contains a piece of his brain.

Science Museum Group. Object number A667469

Fig. 88
Leonardo da Vinci, Anatomical drawing of the muscles of the shoulder and arm, pen and ink on paper, 1510–11.

Wellcome Collection

Fig. 89
'Such the Vultures Love',
students posing with
a dissected cadaver,
photograph, 1910–15.

At the same time, educators were concerned with how dissection potentially affected students psychologically. In the late 18th century, William Hunter declared that 'Anatomy is the Basis of Surgery, it informs the Head, guides the hand, and familiarises the heart to a kind of necessary Inhumanity'.[10] Increasingly, dissection was seen as a 'communal rite of passage' for students, as a transformative and emotionally dangerous time. As one student observed, dissecting 'renders our hearts liable to be corrupted and hardened, renders the sensibilities Callous, [and] brutalises the feelings'. This danger of moral corruption was used as an argument against admitting women to medical studies. Female students who began to enter medical degree programs in the 19th century were careful to stress how the work with dead bodies 'strengthened your womanly feeling, your reverence for the Divine'.[11] The shared experience of dissection was often commemorated with staged photographs in which groups of students posed with cadavers and skeletons (fig. 89). Frequently accompanied by crude mottos ('Such the Vultures Love'), these images rather seemed to confirm the

public's concerns about medical students' lack of respect for the dead.

Questions surrounding the appropriate place of dissection in medical studies continue to the present day. Medical educators explore the possibilities of new synthetic materials and virtual environments to simulate anatomical structures and surgical procedures (fig. 90). However, many medical schools retain an element of training with actual dead bodies.

In the UK, the legal ambiguities that once made bodysnatching a lucrative business have been resolved and the use of corpses is now highly controlled. In 2004 the Human Tissue Act was introduced, which regulates 'the removal, storage, use and disposal of human bodies, organs and tissue'.[12] The 'sacred rituals' that made public dissections more palatable in the past (fig. 91) have been replaced by memorial services for body donors. These special celebrations acknowledge those who have given their remains to serve the improvement of medical knowledge and medical practice in the future.

Fig. 90
Digital technology, such as
this life-sized body imaging
scan, is now frequently
used in medical education.

Anatomage

Fig. 91
J. C. Stadler, after A. Pugin,
*Theatre of Anatomy,
Cambridge*, aquatint, 1815.

Science Museum Group.
Object number 1982-575

A.Pugin del.t   J.C.Stadler sculp.t

THEATRE OF ANATOMY.

London, Pub. Nov.r 1.1815, at 101 Strand, for R. Ackermann's History of Cambridge.

4
THE CHILD IN THE IRON LUNG
POLIO, PUBLIC HEALTH AND PROTECTION

NATASHA McENROE

Fig. 92
Smith-Clarke cabinet respirator or 'iron lung' (detail), Cape Engineering Company, UK, 1953.

Science Museum Group.
Object number 1990-395

The story of polio is a story of many things – of a peculiarly 20th-century disease, of a frenzied search for a vaccine, of the birth of modern fundraising, of the technology used to treat a condition with no cure. Perhaps most of all it is a story that centres around the perception of children and of the fear that can surround them as they face medical treatment. A successful vaccine for polio was discovered in 1955 and it is likely to become the next disease to be eradicated internationally: 2017 saw only 17 new infections worldwide. For the first half of the 20th century, however, the threat of polio struck terror in the hearts of parents and children alike and it was the most feared of all childhood diseases. Because many of the people affected during polio epidemics showed few or no symptoms or effects, the catastrophic damage the disease could inflict on a minority of people seemed even more random. The horror of healthy children being snatched out of ordinary childhoods and condemned to spend a lifetime in callipers or leg irons (fig. 93) or using a wheelchair appeared to imbue the disease with an especially vindictive personality, a personification that was encouraged by vast fundraising campaigns aimed at discovering a vaccine.

Perhaps the most terrifying effect of polio was when it impacted upon the respiratory system. Slow death by suffocation whilst fully conscious, as the muscles needed for breathing became paralysed, is the stuff of nightmares. Paradoxically, what was for some years the key treatment for respiratory paralysis – being placed inside the vast metal giant of the iron lung – is also viewed as nightmarish to us today. But at the time polio was at its most active – in the early 20th century in the USA, post-Second World War in the UK and Europe, and in the 1960s in Russia – iron lungs were a complex symbol in the minds of the parents, medical practitioners and children who came into contact with them, representing both fear and hope. They certainly saved countless lives by 'breathing' for patients with paralysed chests and often made patients feel considerably better (figs 92, 94 and 95). Today, iron lungs and other apparatus historically used to treat the effects of polio are kept in museum collections, a context in which close examination of these objects allows us to imagine a little of the polio patient's experience.

The virus Poliomyelitis (from 'polios' meaning grey and 'myelos' meaning spinal cord) was named in 1847 by the pathologist Adolph Kussmaul. Also known as 'Infantile Paralysis', polio can affect adults but is primarily a disease of young children. The infection causes the spinal cord to become inflamed and, in severe cases, kills the body's motor neurons, which control the muscles. The link between polio and the spinal cord was announced at the 10th International Medical Congress in Berlin in 1890, and it was proved subsequently that polio was a contagious disease, spread by personal contact and, crucially, by people who showed no symptoms during an incubation period, or because they were only lightly affected. Although images from ancient Egypt show people with wasted limbs and the characteristic 'drop foot' often found in people who have been affected by polio, sudden paralysis of infants was not recorded until the 18th century. The virus is transmitted by the faecal to oral route, usually via unwashed hands, which may explain partly why polio was essentially a problem caused by improved modern living conditions. In a less sanitised environment, babies and children would have been exposed to the virus continually and therefore might have built up an immunity in infancy. The first significant outbreak in Britain was in Nottinghamshire in 1835, followed by smaller outbreaks across Europe and the USA. Following this, polio epidemics were referred to by location and date – each one representing a litany of sorrow and fear. Vermont in 1894 was the first wide outbreak in America and an epidemic in Sweden in 1905 had upwards of 1,000 cases. It was at this time that the disease appears to have become much more dangerous, probably due to the virus mutating, and began to attack adults as well as children. The 1917 epidemic in New York City involved over 9,000 cases and 2,400 deaths. Over time, the epidemics were to increase in size and ferocity, often appearing in the summer months, until the discovery of a successful vaccine in 1955.

Amongst all the risks faced by small children in a world without effective vaccines, polio stood out as the most feared. Yet it was not even the biggest threat children faced: compared to the Spanish Flu Pandemic of 1918–19, the death rate was tiny. The killer tuberculosis was far more ubiquitous throughout history, including during the first half of the 20th century, and especially among the poorer classes. Even the common childhood diseases of measles, diphtheria and whooping cough caused more loss of life than polio; a diphtheria epidemic, for example, had a 20 per cent death rate in those who contracted it. Yet polio seems to have held a special place of terror in the hearts of parents, and a lack of understanding about the virus added to the fear they felt. Teething in babies was blamed as a possible cause, as were cats, insects and various sorts of food. More accurately, in the summer months when epidemics were at their height, communal swimming baths were blamed for the spread of the disease and temporarily closed wherever polio was found.

The fear experienced by parents at the potential threat of polio increased a thousandfold if their

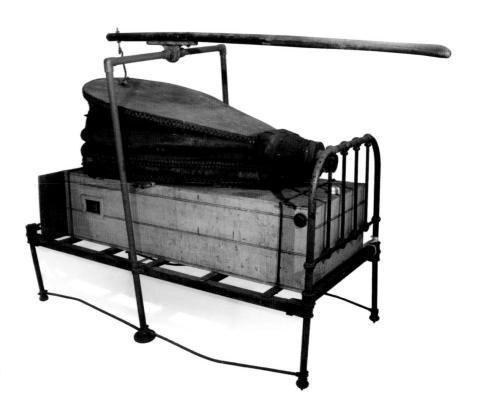

Fig. 93
Child's short leg iron for weak ankles, with shoe attached, from the Lord Mayor Treloar Orthopaedic Hospital, Hampshire, UK, 1960–80.
Science Museum Group. Object number 2002-331

Fig. 94
Custom-built iron lung consisting of a wooden chamber fixed to an iron bedstead and ventilated by hand-cranked leather forge bellows mounted on top, Cardiff, Wales, 1941–50.
Science Museum Group. Object number 1992-67

Fig. 95
Staff nurses with a patient at the Lord Mayor Treloar Cripples' Hospital and College, late 1930s. Photographs such as these inspired Lord Nuffield's gift of iron lungs to hospitals all over the British Empire in 1938.
Science Museum Group. Object number 2002-380

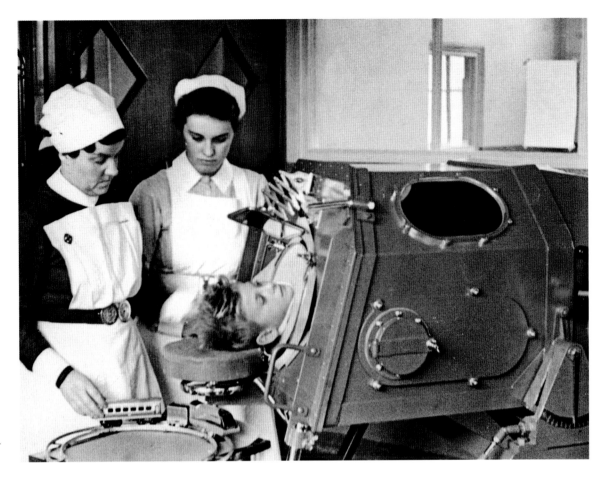

child actually contracted the virus. But what of the child patients themselves? Examining and interpreting the view of a child is always a challenge for a historian. Memories fade and change for all children, and during a period of trauma, they are often lost altogether. The confusion of memories is summed up by one (now-adult) who recalls their experience of having polio aged six:

I do remember being carried into the hospital, having a spinal tap to confirm the diagnosis, and either dreaming or hallucinating that a flock of angels was having a picnic on the empty bed across the double room where I spent the first night of my hospitalization.[1]

However, many personal testimonies have been collected from both American and British adults, who contracted the virus either as children or as adults, and from these we can draw conclusions when a frequent theme is identified. Perhaps the most common remark from these now-adult child patients was the almost complete lack of information given to them either about the virus or their proposed treatment. Tests that included lumbar punctures or spinal taps to draw fluid from the spinal column would need to be forcibly administered, often without warning or explanation to the child. Ann Stevens, a British child who contracted polio whilst her army father was posted in Benghazi in Libya and was evacuated back to a British hospital, notes that, en route, no one explained to her where she was going or why she had left Benghazi.[2] This remark from another child in 1955 is typical:

You never really knew what was going on. I don't think that I even knew what the first surgery was for. They didn't tell you, you didn't dare ask.[3]

It is understandable, perhaps, that faced with what was, quite literally, a life or death situation with few effective treatments available, medical practitioners did not pause to explain matters to the sick child. Additionally, the study of childhood trauma was still in its infancy – and indeed, the connection between healthy minds and bodies is still not fully understood today.

Understanding the effects of separation anxiety and maternal deprivation, caused by children being suddenly separated from their key caregiver, really only gathered pace in the decades following the Second World War, a little late for the many childhood polio patients prior to the launch of the vaccine. As with many infectious diseases, complete isolation of the patient was the primary method of protecting the health of the wider community, and parents of children with polio would have no or very limited contact with their child, especially in the earlier stages of the disease. Later, long weeks or even years could pass while the child remained in hospital, and many families' circumstances meant they were not able to visit frequently.

Apart from the very real threat of death, the dreaded result of an attack of polio was to cause wasted muscles and twisted limbs. As muscles work in pairs, if one muscle is paralysed then its partner has nothing to work against. Therefore, if a patient is young, the still-growing bone can be twisted out of shape by the single, healthy muscle. Treatment for this was like the treatment of tuberculosis of the bone in children – the affected area, the entire body if necessary, would be encased in plaster to prevent the bone from growing out of alignment. One woman who contracted polio when she was 17 remembers coming around from the anaesthetic following surgery:

> I found myself in a body cast that extended from my shoulders all the way down to my left foot. Only my right leg and foot weren't covered by the cast. Before the surgery, no one had told me that I'd be in such a large cast, and when I saw it, I remember being so depressed that all I could do was cry.[4]

If no surgery was required, it was possible to cast the child in plaster whilst they were conscious. Sharon Kimball was nine when she contracted the virus in 1953:

> I remember that they strapped my head in this contraption and dangled me off the floor during the entire process of fitting and moulding that cast. It took about three hours and was gruelling and exhausting.[5]

As well as body casts, callipers, braces and leg irons were fitted on children's weakened limbs and spines both in an attempt to correct the twisted bones and also to give support when the patient attempted to walk independently again.

Most distressing to both witness and experience was when the virus led to bulbar paralysis, which causes the cessation of all the body's control of the respiratory system. Suffocation was the cause of 10 to 15 per cent of deaths from polio, during the acute phase of the disease. Were it not for this (often temporary) inability to breathe independently, the patient might well have made a full recovery. Creating a method whereby the patient could be kept breathing by artificial means became a priority for scientists. As early as 1876, a Parisian, Dr Eugène Woillez (1811–1882), was awarded a silver medal at the Le Havre Exhibition of Life Saving Equipment for his 'spirophone'. This consisted of a large wooden box with incorporated leather bellows designed to fit around the chest of a patient with respiratory difficulties.

In 1918 South African doctors were experimenting with similar large wooden boxes, in which the patient could lie with only their heads and legs emerging from the box, with the waist and neck sealed with clay. These precursors to the so called 'iron lung' had limited success, due largely to problems with creating the necessary airtight conditions. Without this, the air pressure inside the box would not be strong enough to physically manipulate the chest in order to allow the patient to breathe.

Although earlier versions of the iron lung had been attempted, in 1927–28 the industrial hygienist Philip Drinker and the physiologist Louis Shaw from the School of Public Health at Harvard trialled what became the most popular version of the artificial respirator among medical practitioners (fig. 96). Keeping the idea of sealing off the head and neck from the rest of the body by an airtight collar, the iron lung was a metal air-pressured chamber in which the entire body was placed (apart from the head). Drinker was ostensibly inspired by watching a colleague's breathing experiments on cats, which involved the use of a rubber collar. In Drinker's iron lung, a motor was located under the chamber, which worked a lever operating a large rubber diaphragm at the end of the chamber. The movement of the rubber diaphragm in turn created and released negative pressure as the pressure of the air inside the iron lung was lower than in the

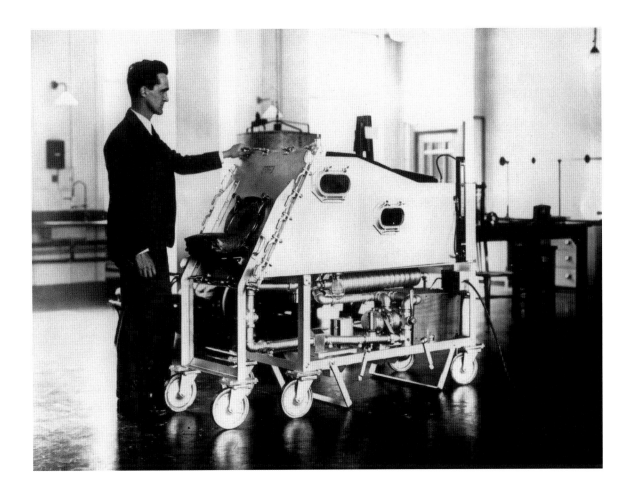

Fig. 96
Philip Drinker with an
iron lung, photograph,
Daily Herald Archive,
c. 1928.

room outside. This caused the patient's chest to inflate and relax, drawing in and expelling air as it did so. The patient was slid into the chamber, lying on their back on a mattress with only their head outside, resting on a small tray at the top end of the bed. Even if the patient's limbs were not all paralysed, they would only be able to make small movements due to the size of the chamber, and they relied on assistance to change position or turn over. The fact that children's bodies tended to be small was an advantage for nurses, as they were easier to handle and move.

The Drinker iron lung (fig. 97) was first used to treat a young girl in Boston Children's Hospital in 1928. Unconscious and with a blue face due to lack of oxygen, the child was placed in the machine by staff (according to Drinker's sister in later accounts, it was Drinker himself who switched on the machine).[6] Sadly, the little girl died – but the next patient, a young man, did not. Alternatives to the Drinker model included the Both portable cabinet respirator, invented by Edward Both in 1937 (fig. 98).

Treatment with an iron lung usually meant being confined to it day and night, especially in the early stages of the disease. Later, some patients could tolerate being out of the iron lung for short periods of time – at least long enough to use the lavatory – and although the hope was that the patient would recover enough to no longer need the machine within a few weeks, in some cases they were required for years or even decades. For those unable to breathe outside the chamber, entry ports were placed at the edges of the machine and bedpans were brought in and out through these apertures. The ports could also be used for washing and for regular repositioning of the body to prevent bedsores. Although the chamber of the iron lung was designed to be at a height to make nursing easier, this remained an awkward undertaking. The high levels of focused care required by polio patients resembled in many ways the nursing in a modern intensive care unit.

Patients who were able to cope outside the iron lung for short periods of time, or even hours, were at an advantage during what all medical staff feared – a power cut. All mechanically operated respiratory machines were fitted with a manual handle, so they could be operated by hand in the event of a power failure, and hospital

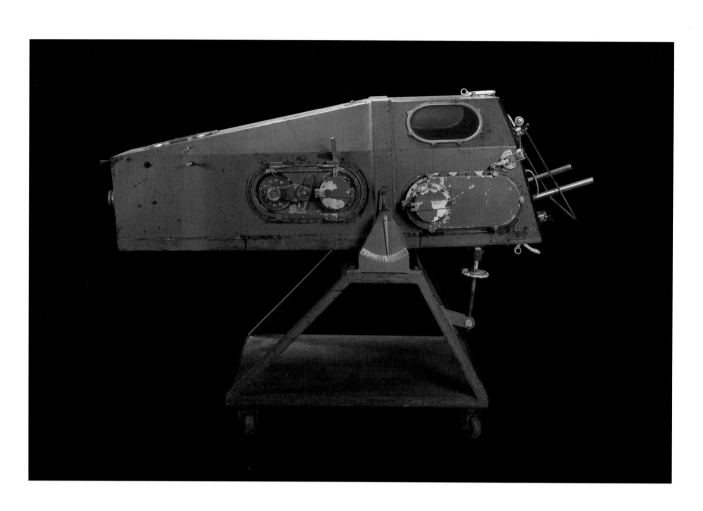

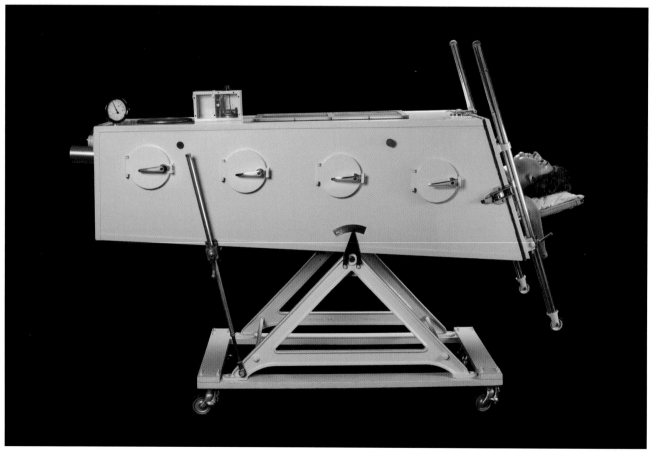

Fig. 97
Drinker-type iron lung
respirator, London,
1930–39. Manufactured
by Siebe Gorman and Co.

Science Museum Group.
Object number 1982-1449

Fig. 98
Both-type iron lung,
London, 1950–55.
Manufactured by D.
and J. Fowler, presented
by Lord Nuffield to the
Memorial Hospital,
Darlington.

Science Museum Group.
Object number A683097

staff were trained to do this. An adult patient, Anne McLaughlin, described how, in 1955, the hospital she was in had invested in 27 petrol-fuelled generators as back up, specifically to keep the iron lungs running for fear of the power failing during a summer thunder storm.[7] Michael Davis, in hospital during an epidemic in Kentucky in 1944 when he was aged 13, recalled that when 'the continuous wheezing and clanking of the iron lungs' stopped, this was a signal for immediate action.[8]

The best outcome for treatment of polio with an iron lung was that after use during the worst period of bulbar paralysis, an independent life could be resumed. Treatment using the machines was not easy to administer owing to difficulties in measuring lung capacity: attempts were rather rudimentary, with medical staff simply counting as the patient breathed out and adjusting the rate and pressure of the machine's 'breathing' accordingly. Another equally approximate method was to monitor the mental capacity of the patient, since ventilation controls not only oxygen levels in the blood but also carbon dioxide levels, and changes in either of these two things can impact dramatically on the individual's ability to think clearly. Predictably, both lung capacity and mental capacity must have been near impossible to monitor in the case of a distressed child.

The child in the iron lung was truly isolated, not only from their familiar life but also, physically, from the external environment. Their body was in effect divided into two – most of it encapsulated in the metal chamber with only their head outside. For those who were conscious, mirrors were placed above the headrest, so the patient was able to see what was happening in the room around them. This mirror could double up as a book support and was frequently used as such by patients who were old enough to read, but they relied on nurses or an ambulant patient to turn the pages for them. The experience of a healthy person suddenly having their breathing performed by machine is described by the journalist Drew Pearson in an episode of the news programme *Washington Merry-Go-Round* in about 1956, when he tried the iron lung as part of an experiment for a short film.[9] Once inside the iron lung, Pearson explained how tight and uncomfortable the neck collar was, whilst he was instructed by the two nurses supervising him to regulate

his breathing to match that of the machine. Speech and swallowing was only possible during the outward breath, rendering the iron lung even more alien. Of course, Pearson was a healthy man, but his description of the sense of isolation produced by being in the contraption is striking. He said that he felt:

> … away in a world all your own, with nothing to look at except this little mirror, and nothing else in the world except that.[10]

There are very few personal testimonies from children with experience of an iron lung, probably due to fading memories or a dislike of looking back to a traumatic time. One that does exist is by Marilynne Rogers, who, having used an iron lung in childhood, continued to need one in order to sleep until her death, decades later. Marilynne's account is unusual in that she trusted the nursing staff, mainly because her cousin was a nurse. Perhaps that is why her description of the experience is not entirely negative. She contracted polio at the age of nine in 1949:

> I remember they opened the respirator; it seemed really huge and they laid me on the tray with a mattress on it, and then they slid me though the hole at the front of the big roller part. Then they closed up the collar and told me to really relax. They told me I'd feel much better, and I did. I could breathe more easily.[11]

Marilynne's account has the usual weakness of childhood memoirs – that caused simply by the passage of time. She knows from her medical records that she was in and out of the iron lung for weeks but has no memory of the experience beyond the description above.[12]

Aside from the trauma and discomfort experienced by a user of the iron lung, it was also beset with clinical problems. Although its invention was heralded as the key to keeping polio sufferers alive during temporary periods of bulbar paralysis, patients who used it continued to face a high mortality rate, often from lung infections or broncho-pneumonia caused by their inability to cough or clear excess fluid from the lungs. Taking the occasional deep breath – or even sighing – is important for healthy lung function and was impossible with machine-regulated respiration. Practitioners were aware of this and tried to overcome the problem in several ways, such as by adapting the iron lung's motor to allow for coughing or sighing, or by draining fluid from lungs manually with tubes. Other forms of treatment were used in alternation with the iron lung – the rocking bed, for example, was an ordinary bed mounted on a hinge to allow for a see-saw motion. This caused the patient's own diaphragm to move with gravity, assisting breathing. People adjusted surprisingly quickly to the continual motion and were even able to sleep in them. More desperate measures included tracheotomy – cutting a hole in the throat of the patient and pumping air into the lungs directly. Various types of fitted respirators were used too, encircling the patient's chest, squeezing and releasing to cause breathing. These methods of avoiding the build-up of fluid in the lungs had the advantage of allowing the patient to be more ambulant than when in the iron lung. Different combinations of these various methods were used in the treatment of polio and its associated respiratory difficulties.

None of these methods – even the drastic and alarming tracheotomy – seem to have frightened patients to the same degree as the iron lung. The singer Joni Mitchell, who had polio at the age of nine in 1952 in Saskatoon, Canada, has spoken of her childhood fear of iron lungs whilst hospitalised, especially at night:

> … not so much in the daytime because the halls were full of activity, but at night, the sound of the iron lungs – that wheezing breathing – it was a terrible sound and we all dreaded the possibility that we could end up in one of those cans.[13]

This dread is echoed throughout the personal testimonies of many children. Dr Robert M. Eiben, based in Cleveland, Ohio, writes, 'It is unlikely there was ever a polio patient who was not fearful of the iron lung.'[14] Dr Eiben also states that if a patient panicked badly when being placed in one, they would be immediately taken out. He does not state if this was the case for children as well as adults, but one hopes that it was.

The iron lung had no fiercer opponent than the self-styled 'Sister' Elizabeth Kenny, an Australian woman who had acted as a volunteer nurse during the First World War and later claimed that working with injured soldiers had

enabled her to create her controversial methods for dealing with the effects of polio. She opened a clinic in Townsville, Australia, in 1933, and following a move to the USA she opened the Sister Kenny Institute in Minneapolis in 1942, dedicated to helping polio patients. Sister Kenny fought against both plaster-casting and iron lungs, rightly believing that weakened muscles would only grow weaker if immobilised. Her treatment concentrated instead on continuous movement, placing packs of heated blankets over the affected areas and painting with hot wax, all with the aim of keeping muscles pliable. Unlike many practitioners, Kenny did explain her methods to the children she was treating, though she was described as having a robust 'no-nonsense' approach. Unfortunately, this attitude carried over to her dealings with funding bodies, including the mighty National Foundation for Infantile Paralysis (NFIP), which funded her initially and then withdrew its support when it found her too difficult to work with. She treated polio patients for many years and, although her methods were controversial, they spread throughout the world.

Kenny's methods of manipulation of limbs was also experienced by the British singer-songwriter Ian Dury (1942–2000) (fig. 99). Dury contracted the polio virus aged seven in 1949 – he believed from the swimming baths in Southend – and was sent to Black Notley Hospital in Essex. Here staff used painful physiotherapy and manipulation to straighten children's limbs. Dury remembers:

> It was called the screaming ward and you could hear people screaming on the way there, and it was you when you were there, and you could hear the others on the way back.[15]

Dury was eventually placed in the Chailey Heritage school in Sussex. Chailey was ground-breaking in its 'tough love' approach: they believed that if a child fell over, they had to get up again themselves. On an episode of the television chat show *Parkinson* in 1981, Dury explained that when he was at Chailey, the school was 'a bit of a roughhouse' as it was transitioning from being a charitable institution to part of the National Health Service, which had been launched in 1948.[16] Although methods were aimed at encouraging independence, his

Fig. 99
Ian Dury performing live
on stage, 1977.

experience was brutal and this is sometimes
reflected in his lyrics:

Hey, hey, take me away
I hate waking up in this place
There's nutters in here who whistle and cheer …

When I get better, when I get strong
Will I be alright in the head?
They're making me well, if they're caring for me
Why do they boot me and punch me?
Why do they bash me and crunch me?[17]

The NFIP in America was responsible for vast
amounts of research into polio. Set up in 1938,
it had raised $630 million by 1962. The 'March
of Dimes' remains one of the largest and most
successful funding drives of all time and, in some
ways, was where modern fundraising began,
with its use of radio and short films and the
involvement of celebrities. It was enormously
influential. The science writer Gareth Williams
cites the intense fear of polio in America – even
greater than in Britain and Europe – as a direct
result of the March of Dimes's success, which
manipulated an already concerned American
public into charitable giving. The aim of the
campaign was to fund research to find a
vaccine,[18] and the concept was simple – donating
a dime (ten cents) was something that even a
child was able to do. Initially contributions were
posted to the White House, followed by local
collecting booths and high-profile fundraising
balls. The British Polio Fellowship was, as the
name suggests, a very different organisation
to its American counterpart. The Fellowship
concentrated largely on supporting people
affected by polio and saw fundraising as
only a secondary mission. These, and other
organisations, created short films that offer
the historian today an interesting insight into
polio's place in our past from a different point
of view.

Archival film relating to polio falls into two
distinct categories, both of which raise different
problems as historic sources. Many short films
were created for use purely within the medical
profession, and these frequently have little or
no contextual information. But most of the short
films about polio were created for fundraising
purposes and, as such, often combine pathos
with hope. One film, *Hospital School*,[19] depicts
the Lord Mayor Treloar Cripples' Hospital in
1945 (fig. 100). Founded in 1908, it was initially
for children with tuberculosis of the bone but
admitted polio patients from the 1920s onwards.
The ten-minute film shows child patients entirely
happy and serene, even when undergoing a body
plaster-casting treatment – a truly unpleasant
experience. As with fundraising promotions
today, the correct balance had to be found so
as to tug at the public's heartstrings, but not
distress them so much that they switched off.
The best hook for fundraising in both America
and Britain was that extremely well-known polio
patient, the President of the United States of
America, Franklin D. Roosevelt. Roosevelt
contracted polio as an adult whilst holidaying on
Campobello Island off New Brunswick, Canada,
in 1921. Despite largely concealing his disability
(he was unable to walk unaided but only one
photograph of him exists using a wheelchair)
he remained a powerful role model for children
and an excellent tool for fundraising. Following
his death in 1945, his widow, Eleanor, narrated,
along with the actor Michael Redgrave, the short
film *His Fighting Chance*,[20] which presented

Roosevelt as one of the great men of our time, and one who had won the battle against polio. The film features 'little Johnny Green', a paralysed toddler who, by the end of the film, was able to lift his own head from his bed.

Perhaps the most famous of the fundraisers was the radio programme produced by the March of Dimes in 1943. *The Crippler* featured the actor Raymond Massey as the voice of the villainous Infantile Paralysis – 'the Crippler'. It combined perfectly the anxiety that any child can be 'touched' by polio with the optimism of the work carried out by the NFIP:

All I have to do is reach out … like this …. And when I touch a child, there is pain, there is suffering … I can do it any time, to anybody, without hindrance. Well, *almost* without hindrance. Lately there has been some interfering with my work.[21]

The character of Polio itself is furious to observe a small girl in hospital learning to walk again – thanks to the NFIP. 'Where did she get all that courage and hope?' snarls the Crippler.

All methods of fundraising for children with disabilities – or indeed any form of fundraising for children – rely heavily on the image of the child. Attractive, brave and, above all, visibly in need, real child polio patients were used as models by the campaigns, their party clothes worn with the metal leg irons or callipers that allowed them to walk. The child with some form of walking aid was a ubiquitous trope of fundraising at the time of the polio epidemics, not just an image for the short films and posters but also for collecting boxes placed on every high street. Lifelike models of children with crutches, children with leg irons, looking piteous yet brave, were a frequent sight on British street corners (fig. 101). These collecting boxes were a method of fundraising that continued up until the late 1970s, long after polio ceased to be a threat, adapted by other charities such as the Spastics Society (now Scope.) By the 1980s, the image of a begging child went against the spirit of independence for people with disabilities that charitable organisations were trying to foster. One collecting box that marks a curious transition of this period of social change is in the Science

Fig. 103
Arm splint for a young girl
recovering from polio,
Dartford, UK, 1942–48.

Science Museum Group.
Object number 1989-832/4

Museum collections (fig. 102). Unlike previous models, 'Debra' is not a lifelike child but rather a cartoon character – although still wearing the familiar callipers on her leg.[22]

Following a number of extensive trials, a vaccine to protect against polio was announced in 1955. This caused a media frenzy and is now recognised as the second most successful vaccination programme of all time, beaten only by the smallpox campaign. Although pockets of polio still exist in Pakistan and Afghanistan – owing to a combination of remote rural communities, regional conflict and anti-vaccination propaganda – the aim to make the world polio-free now seems achievable. Polio is no longer present in the Western world, and even those whose bodies show the effects of polio are often not associated with it. A prime example of this is the celebrity cook Mary Berry, who had polio at the age of 13 for three months and whose hand and lower arm are visibly affected.[23] Despite the swift disappearance of polio from the West, however, iron lungs continued to be used for some years for other respiratory conditions, such as kypho-scoliosis, myasthenia gravis, muscular dystrophy and neurological issues.

Roisin Tierney worked as the Departmental Sister in the Intensive Therapy Unit at St Thomas' Hospital and then as the Nursing Officer in the Supra Regional Respiratory Unit, Phipps Ward, South Western Hospital under the auspices St Thomas' Hospital from 1965 to 1983, later becoming a Sister in the Royal Household. She frequently worked with patients in iron lungs, of whom the majority had had polio in the years before the vaccine became available, but also other patients with a range of restrictive respiratory problems requiring non-invasive ventilation. Unlike other testimonies, she felt that iron lungs were not usually a source of fear for her patients. She writes:

> The majority of patients I cared for were used to iron lungs and when unwell were relieved to be able to be placed within one. Those who were unfamiliar with iron lungs would probably have been transferred as emergencies and were usually too unwell on arrival to notice what was happening. The iron lung size is daunting and the prospect of being 'shut up' was frightening to some of the 'novices'.[24]

Tierney's memories demonstrate how the role of the nurse was still crucial in caring for adult non-polio patients as well as post-polio patients.

> Each new member of staff was asked to go into an iron lung so they would be familiar with the feeling of having their breathing totally controlled ... If a patient's breathing was not synchronised with the machine either the breathing pattern was dictated by a nurse watching the dial and saying 'In', 'Out' until the breathing was in time with the machine, or the pressure could be turned up to the point that the machine had totally taken over, at which time the pressure could be lowered to the normal.[25]

What can museum collections tell us about the experience of the child in the iron lung? How could small children be communicated with, when ill and distressed at being separated from their parents? The polio-related objects and their records at the Science Museum share numerous, complex stories – and by combining the objects themselves with personal testimonies, a clearer picture begins to emerge. We can see this demonstrated in a group of four arm splints that one British woman, Mrs Blackstone, used when she was infected with polio as a child between 1942 and 1948 (fig. 103). The splints show her gradual recovery, with each one made of a lighter material than the one that preceded it. They are a record of a child's improvement in health following her contraction of the polio virus: since muscle is strengthened by use, weights were added to splints so as to increase the work done by the muscle during everyday movement. As the muscle strengthened, less weight was required to build it. In the paperwork created at the time of the donation of the splints to the Museum in 1989, which included a personal testimony from their user, we learn that Mrs Blackstone became distressed at the sight of the splints more than 40 years after she used them:

> The splints were found in a cupboard in my former house at Erith when my mother recently moved out. The sight of them after so many years was a somewhat disturbing one.[26]

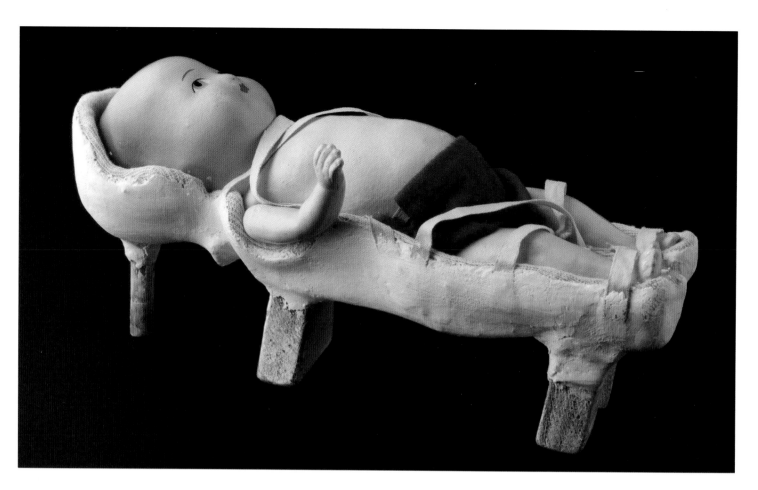

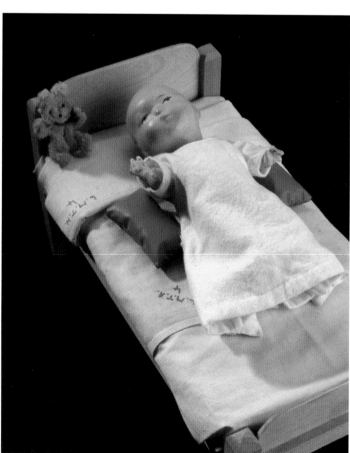

Fig. 104 (above)
Ceramic infant doll, encased within a full body-length supported plaster cast, probably used to demonstrate to child patients their prospective treatment for scoliosis, polio or skeletal TB, from the Lord Mayor Treloar Cripples' Hospital and College, Hampshire, UK, 1930–50.

Science Museum Group. Object number 2002-360

Fig. 105 (left)
Ceramic infant doll on a model bed, probably used to demonstrate to child patients their prospective early treatment for poliomyelitis, complete with supportive pillows and pads, bedding, small teddy bear and descriptive label, from the Lord Mayor Treloar Cripples' Hospital and College, Hampshire, UK, 1930–50.

Science Museum Group. Object number 2002-359

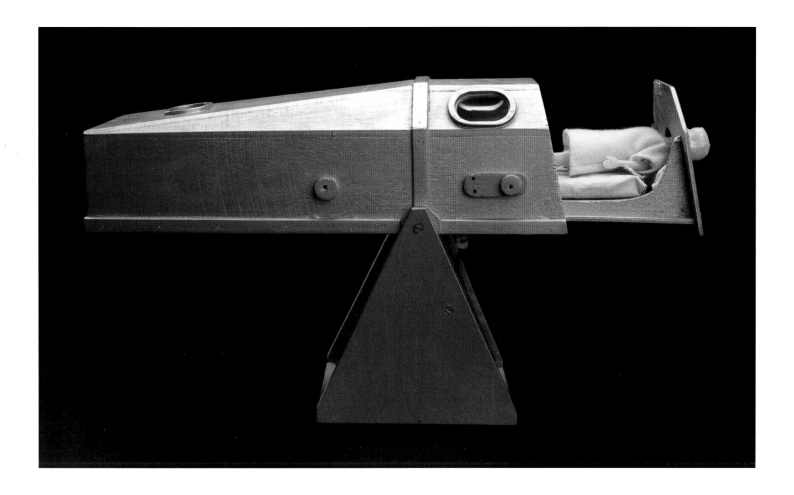

Fig. 106
Wooden model of an
iron lung, with plastic
infant doll and bedding
enclosed, probably used
to demonstrate to child
patients their prospective
treatment for poliomyelitis,
from the Lord Mayor Treloar
Cripples' Hospital and
College, Hampshire, UK,
1930–50.

Science Museum Group.
Object number 2002-363

The most troubling element in the personal testimonies of previous child patients is their distress at the lack of communication with them from those in authority about their condition and its treatment, at a time when their parents were absent – itself a psychologically damaging situation. This may have been partly because medics were often working at crisis level during medical emergencies and simply did not have time to spend talking to their patients, and partly due to a culture that still had its roots in the Victorian age in terms of the importance of children's emotions. To modern eyes, it seems that medical staff failed to provide appropriate care for the children in ignoring their emotional responses to events. However, some Museum objects suggest that some staff were in fact attempting to communicate with children about the drastic procedures they were facing, even if this ambition was not fully realised in the experience of the child. A group of ceramic dolls (figs 104–6) is perhaps the most touching of these objects. They were collected by the Science Museum in 2001 when the Lord Mayor Treloar Orthopaedic Hospital (previously the Lord Mayor Treloar

Cripples' Hospital and College) moved from its original site to new premises.

The dolls are inexpensive toys that might have been owned by any little child, but they are adapted in a variety of ways that form an important part of the medical record. Some of the dolls are set partially or fully in body casts (fig. 104), and some come with their own beds – one even has its own miniature teddy bear (fig. 105). A model iron lung made of light wood and painted grey has a functioning drawer, and upon it a tiny plastic doll is placed (fig. 106). The doll in the iron lung is slightly out of proportion, even for a child in a full-size iron lung (one assumes it was the best-sized doll available) but it serves only to emphasise the small body of the child in the giant machine. Precisely how these dolls were used is unknown, but their aim is clear – to communicate about the experience of the child undergoing treatment for polio with a child audience. We can only hope that the use of these objects, now cared for in a Museum store, made the horrible experience of polio slightly less bewildering for at least one child.

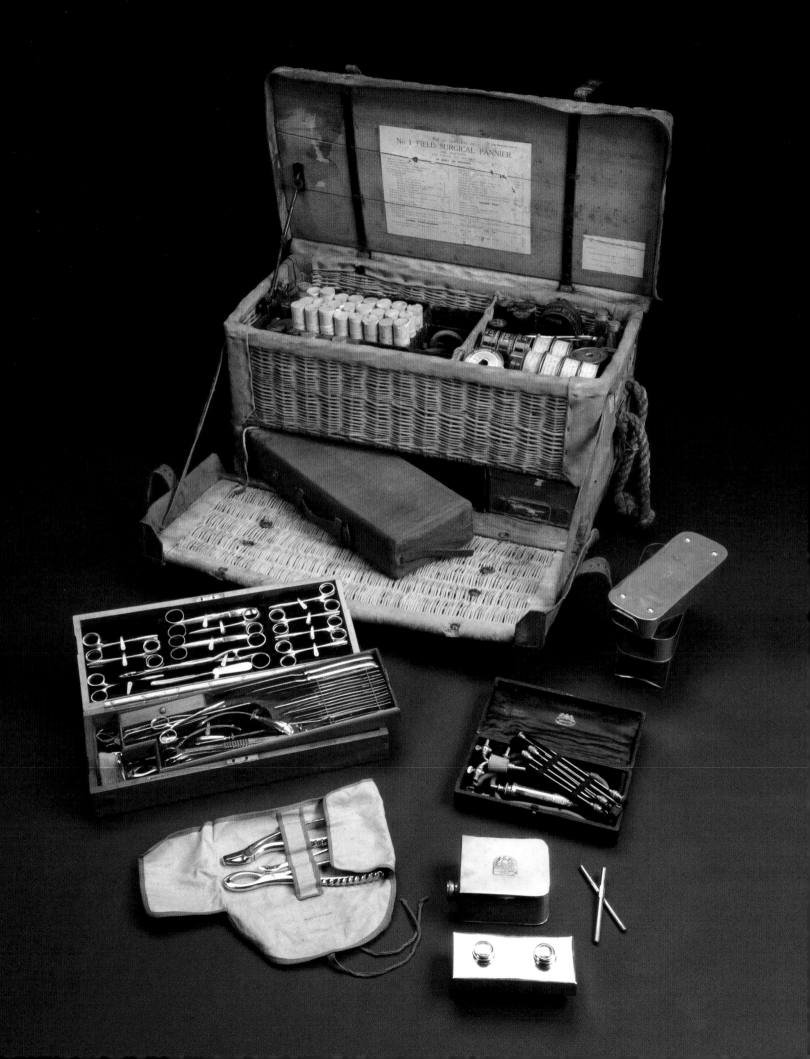

5
CREATED THROUGH CONFLICT
THE DEVELOPMENT OF MILITARY MEDICINE

JACK DAVIES

Medicine and war have always had a turbulent relationship. Though conflict by its very nature destroys bodies, damages health and decimates landscapes, it has also provided the medical profession with bodies, experience and opportunities for research. Unfortunately, developments in artillery have a habit of progressing more quickly than medicine ever has. New weapons create new wounds, a correlation that became increasingly apparent during the 20th century. Further problems were caused by improvements to transport technology, which meant that wars were fought in new lands, each with its own environmental challenges, including the adequate distribution of supplies. These experiences have changed the meaning of military medicine and helped it to flourish into a highly specialised branch of modern medicine.

Surgeons have operated on soldiers for thousands of years. Hippocrates, the so-called 'father of medicine', wrote in 400BC that 'He who wants to be a skilful surgeon must have sufficient experience of military surgery, namely, war surgery', for 'war is the only proper school for a surgeon.'[1] However, in the medieval period, it was barber-surgeons rather than learned practitioners who held the scalpel. From the 1660s onwards, each regiment of the British army had its own surgeon and assistant, but it was not until the Peninsular Wars of 1810–14 that the medical services were organised more formally. From the 1800s surgeons were university-educated, a trend that intensified through further professionalisation. By the time of the First and Second World Wars, the British

army relied heavily on civilian practitioners who volunteered their services (fig. 107). Today, it is usually only highly trained specialists who work in military medical contexts.

This transition was only possible due to the significant advancements in science and medicine of the 1800s. In 1846 the American dentist William Morton demonstrated the use of ether as an anaesthetic while extracting a tooth, allowing surgeons to better manage their patients' pain (fig. 108). In 1867 the surgeon Joseph Lister proved the importance of antiseptics through his use of carbolic spray, a mixture of carbolic acid and water that lowered the risk of infection for surgical patients (fig. 109). In the 1870s the biologists Louis Pasteur and Robert Koch (1843–1910) established germ theory. Prior to this discovery, it was believed that diseases were caused by spontaneous generation and spread through bad smells. Their work proved the presence of disease-causing organisms, prompting improved hygiene practices. These discoveries ensured that more patients than ever before survived their operations, allowing surgeons to complete procedures that were once unthinkable.

The Science Museum's medical collection is based largely around Henry Wellcome's collecting. We have a substantial collection of military medical equipment from a variety of different conflicts that occurred before his death in 1936. These objects spoke to Wellcome's aim to collect the human experience – something to which, for better or worse, war is integral.

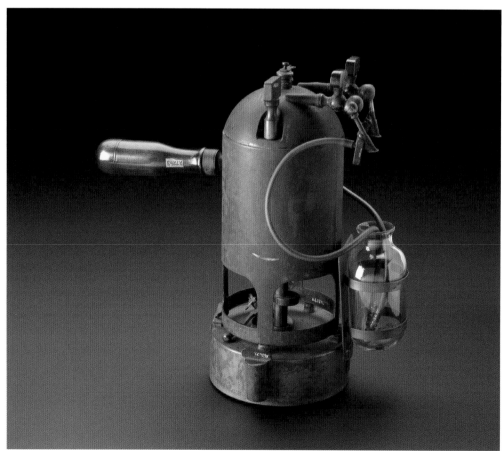

Fig. 108 (above)
Copy of William Morton's
inhaler, 1870–1920.
The first recorded use of
ether as an anaesthetic
was on 16 October 1846,
by Morton, an American
dentist.
Science Museum Group.
Object number A625379

Fig. 109 (left)
Steam carbolic spray,
London, 1871–1900.
Joseph Lister designed this
spray to cover everything
in the operating theatre or
hospital ward with carbolic
acid or phenol vapour.
The aim was to create
an antiseptic environment
during surgery and to
prevent the spread of
disease in the ward.
Science Museum Group.
Object number A43471

Fig. 110
Louis Pasteur,
portrait photograph,
Photographische
Gesellschaft, Berlin,
1910–19.

Science Museum Group.
Object number 1982-1459/57

Fig. 111
Robert Koch,
portrait photograph,
Photographische
Gesellschaft, Berlin,
1910–19.

Science Museum Group.
Object number 1982-1459/36

THE CRIMEAN WAR (1853–56)

The Crimean War saw Britain, Russia and France fighting the Turkish Ottoman Empire. Most of the fighting took place on the Crimean Peninsula during a particularly harsh winter, but the conflict is most famous for its high mortality rates. Of the 60,000 men enlisted in the British army, 21,000 perished, but only 4,500 of them died from their wounds. The rest succumbed to disease or infection.

Britain's lack of recent combat experience (the last major conflict had been in 1815), meant that they were ill equipped to deal with the problems of the Crimean War. There were too few staff and inadequate supplies, resulting in unsanitary military hospitals. Most of the medical care provided during the conflict revolved around infectious diseases rather than surgery, as few soldiers survived long enough after wounding to receive surgical attention. Those who did had bullets and debris removed from wounds and often had limbs amputated. Luckily for them, the Crimean War was the first British military campaign to use anaesthesia close to the battlefield.

The Principal Medical Officer, Dr Hall, went against medical protocol and discouraged the use of chloroform as an anaesthetic. At that time, it was common for doctors to choose for themselves which anaesthetic agent to work with, but Hall insisted that all practitioners working under him use ether over chloroform. Though both were discovered at a similar time, the medical profession leant towards chloroform due to ether's instability and flammability. Despite not being approved by Hall, chloroform was used throughout the conflict. As with most medical supplies in the Crimea, demand far outstripped supply. Edward Mason Wrench, a British army surgeon, wrote that:

> We were practically without medicines, the supply landed at the commencement of the campaign was exhausted, and the reserve had gone to the bottom of the sea in the wreck of the Prince so that in November 1854 even the base hospital at Balaclava was devoid of opium, quinine, ammonia, and indeed of all important drugs.[2]

The lack of drugs available to medical services had a significant impact on the lives, and indeed the deaths, of wounded soldiers. Wrench continued to describe his experiences with the sick in another of his letters:

> I had charge of from 20 to 30 patients, wounded from Inkerman, mixed with cases of cholera, dysentery, and fever. There were no beds ... or proper bedding. The patients lay in their clothes on the floor, which from rain blown in through the open (i.e. broken and unrepaired) windows, and the traffic to and from the open-air latrines, was as muddy as a country road.[3]

It is clear that it was not just drugs and medicines that were in short supply.

The recent invention of the telegraph in the 1830s meant that news of the army's lack of preparedness was quick to reach Britain. The press reported on the lack of supplies and unsanitary conditions inside military hospitals, prompting Mary Seacole's and Florence Nightingale's attempts to care for British soldiers.

Florence Nightingale (1820–1910) is perhaps the best-remembered figure of the Crimean War and is revered as one of Britain's great heroes (figs 112–14). Born in Florence, Italy, she was raised in England. She was highly educated and sought a career in nursing after what she claimed was a call from God. After reading about the poor conditions and inadequate supplies in Crimean War hospitals, she led a group of women to Scutari hospital in Constantinople, Turkey. All of Nightingale's nurses were clothed in the Scutari sash (figs 115 and 116). The conditions in the hospital were depicted in a series of paintings and prints now held in the Science Museum collections (fig. 117). When they arrived, the women were shocked to discover how few supplies were available, how overworked the staff were and how neglected the sanitation and hygiene of the hospital had become.

A staunch believer in miasmic theory, Nightingale thought that diseases were transmitted through foul air. Consequently, she set to work implementing new hygiene practices. Nightingale used her influence to push for improved sanitation, and her voice joined others expressed through

Fig. 112
Parasol owned by Florence Nightingale, brass, ivory, wood and cloth.

Science Museum Group.
Object number A6101

Fig. 113
Silk shawl owned by
Florence Nightingale.

Science Museum Group.
Object number A87224

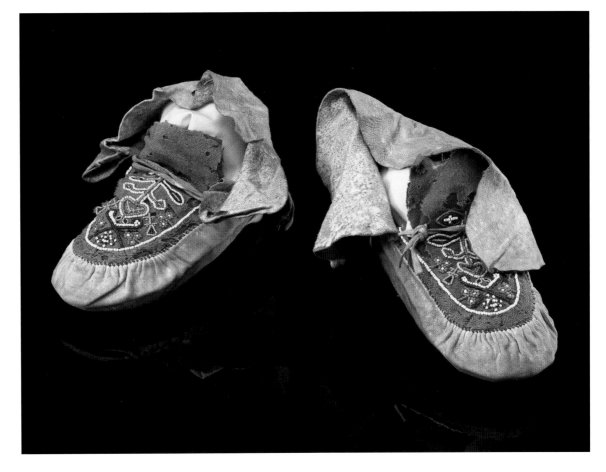

Fig. 114
Pair of leather moccasins,
said to have been worn
by Florence Nightingale,
leather, glass, beads,
cloth, 1850—56.

Science Museum Group.
Object number A96087

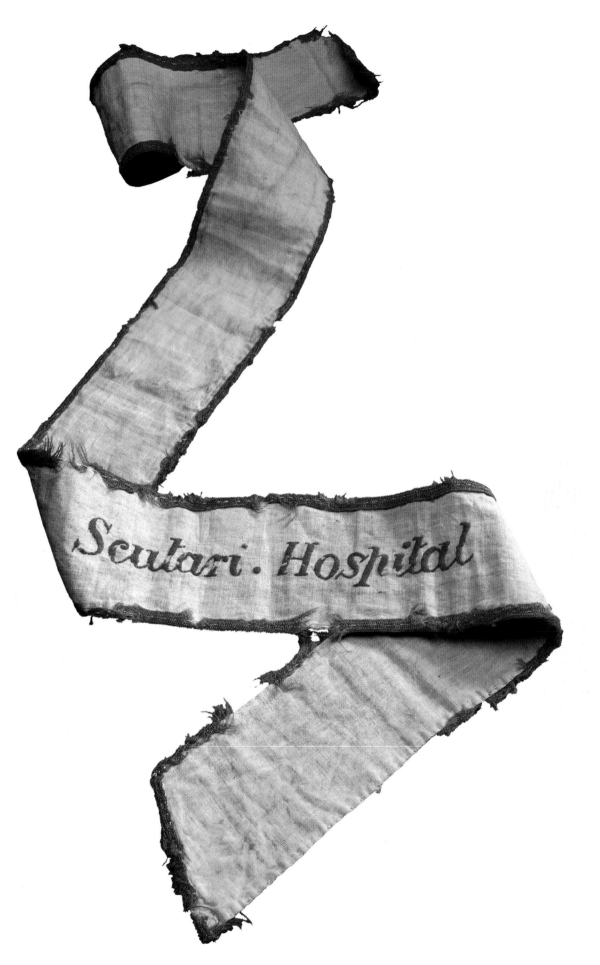

Fig. 115
Nurse's sash for Scutari
Hospital, c. 1855. This
example was owned
by Charlotte Wilsdon.
Science Museum Group.
Object number 1979-352

Fig. 116
Charlotte Wilsdon, a
nurse who worked with
Florence Nightingale on
the Bosphorus and in the
Crimea at Scutari Hospital,
portrait photograph,
1850–59.
Science Museum Group.
Object number 1986-1375/1

Fig. 117
Edmund Walker, after W. Simpson, *One of the Wards of the Hospital at Scutari*, lithograph, published by Colnaghi, 1856.

Science Museum Group.
Object number 1987-199

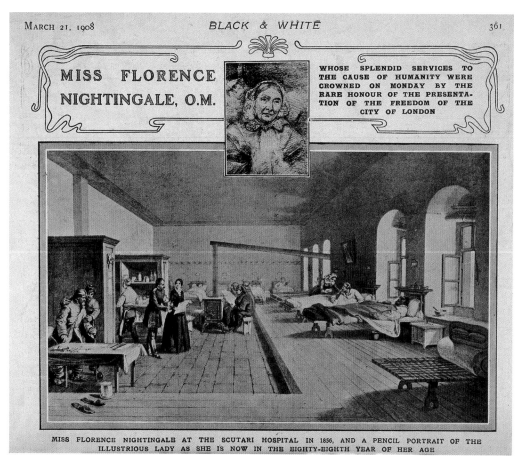

MARCH 21, 1908   BLACK & WHITE   361

MISS FLORENCE NIGHTINGALE, O.M.

WHOSE SPLENDID SERVICES TO THE CAUSE OF HUMANITY WERE CROWNED ON MONDAY BY THE RARE HONOUR OF THE PRESENTATION OF THE FREEDOM OF THE CITY OF LONDON

MISS FLORENCE NIGHTINGALE AT THE SCUTARI HOSPITAL IN 1856, AND A PENCIL PORTRAIT OF THE ILLUSTRIOUS LADY AS SHE IS NOW IN THE EIGHTY-EIGHTH YEAR OF HER AGE

Fig. 118
A print reporting that Nightingale had received the Freedom of the City of London, 1908. The image is a reprint of Walker's lithograph above.

Wellcome Collection.

the British press, eventually resulting in the Sanitary Commission of 1855. This cleared the sewers and improved ventilation in hospitals, thereby reducing the death rate of patients recovering at Scutari.

The Times reported on her activities, making her an overnight sensation in Victorian Britain:

> She is a ministering angel without any exaggeration in these hospitals, and as her slender form glides quietly along each corridor, every poor fellow's face softens with gratitude at the sight of her.[4]

A few weeks later, the *Illustrated London News* christened her as 'The Lady with the Lamp', an identity that was celebrated through many forms of contemporary media.[5] She is often described as an icon of Victorian culture and remains one of the most famous British female military heroes (fig. 118).

Nightingale's celebrity was so great that she recorded a speech for the Light Brigade Relief Fund in 1890, forever immortalising her voice as an historical record on a graphophone wax

cylinder (like the one in fig. 119). Perhaps her biggest achievement was establishing the nursing school at St Thomas' Hospital in London. Her work in the Crimea and in training future generations of nurses improved the reputation of the profession, which prior to the Crimean War was heavily associated with female prostitution.

Nightingale was by no means the only woman to make her mark on the Crimean War. Mary Seacole (1805–1881) (fig. 120), born Mary Jean Grant, also cared for British soldiers in the Crimean. She was born in Kingston, Jamaica, to a Scottish father and Jamaican mother who passed down her knowledge of nursing. Seacole nursed patients in Panama after cholera outbreaks, emphasising the importance of cleanliness and ventilation. Upon hearing of the adverse conditions in hospitals in the Crimea, she tried to join a nursing party. Seacole was one of the many women who were not accepted to travel to the region. Women were rejected for a number of different reasons, but it is likely that Seacole was discriminated against because of her race. Undeterred, she funded her own trip,

spending a night with Nightingale at Scutari hospital before moving on to Balaclava. Lacking adequate building supplies, Seacole collected pieces of wood and metal wherever she could, which later formed the bulk of the building material for her institution, the British Hotel that opened near Balaclava in 1855. Here she fed and cared for a number of British soldiers, providing hospitality rather than medical care. Seacole's philanthropic nature was widely praised in the British press, but her efforts in the Crimea left her in financial ruin.

Nightingale, like Seacole, is often regarded as a British military hero. Both provided care and comfort to Britain's wounded soldiers, and both are examples of women who made a difference in the male-dominated military environment of the 1800s. The Crimean War taught the British army many valuable lessons about health and hygiene, as well as the importance of nursing. It helped to streamline the military medical service and improve its structure. In the aftermath of the conflict, doctors in the British army were brought together under the Medical Staff of the Army Medical Department, while stretcher-bearers and orderlies belonged to the Medical Staff Corps. These were later united to create the Royal Army Medical Corps (RAMC) in 1898. Military nurses were registered in the Army Nursing Service in 1881.

THE SECOND BOER WAR (1899–1902)

The Second Boer War, or South African Conflict, was fought between the British Empire and two different Boer States. Britain did not anticipate much resistance, assuming that the Boers would be unable to match the might of the British army. They were wrong. The Boers struck first by besieging Ladysmith (in Natal and Kimberley in south-east South Africa) and Mafeking (on the border with Botswana), but the British eventually broke the siege, forcing the Boers to employ guerrilla tactics. For two years the Boer soldiers ambushed the British forces and escaped quickly before they could retaliate. The whole campaign was embarrassing for the British government, who eventually resorted to interning Boer civilians in concentration camps and burning their crops to stop food supplies, resulting in large numbers of civilians dying from preventable diseases.

Much like the Crimean War, the Second Boer War was far more difficult for doctors than surgeons. Though army officials had improved their understanding of health and disease and upgraded the sanitary conditions that soldiers were serving in, they still found themselves struggling to cope with the number of sick patients. The Royal Army Medical Corps had only been established 16 months prior to the war and found itself understaffed and ill equipped. Hospitals admitted 22,000 British troops due to wounds caused by enemy forces, while 74,000 contracted enteric fever and dysentery. Typhus was also rife. Though Sir Almroth Wright created a vaccine in 1896, it was not mandatory and only five per cent of soldiers took it. The other 95 per cent refused as they were wary of its side effects: it was known to cause fever and malaise, and consequently typhus remained a problem throughout the war.[6]

It was not just education around vaccination that soldiers lacked: most were still not equipped with adequate knowledge of hygiene and disease prevention. Maintaining sanitary standards usually fell to the Regimental Sanitation Officer, but the post had been abolished shortly before the war, so the responsibility fell to untrained officers instead. This meant that soldiers were often left with poor facilities. Most of the diseases that affected British soldiers in South Africa were related to sanitation. Some soldiers were supplied with charcoal water filters, one of which we hold in the Science Museum collection (fig. 121). The porous nature of charcoal allows it to absorb pollutants, purifying the water that passes through it, making it safer to drink and preventing the transmission of many waterborne diseases. Unfortunately, first-hand accounts from the Boer War suggest that too few water filters were provided.

The lack of fresh water supplies or effective, readily available means of purification meant that soldiers drank from the germ-ridden Modder River, exposing themselves to many different diseases (fig. 122).

There was nothing at all to eat but bully beef and biscuits, and nothing to drink but muddy water. At several points the road was blocked by rain. We were stopped and kept three hours standing in the mud and pouring rain, the men not having either blankets or overcoats.[7]

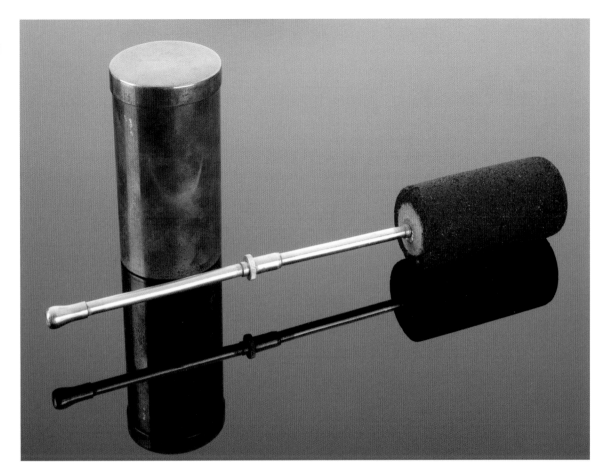

Fig. 121
Portable charcoal filter and brass case, c. 1890–1900. This filter is believed to have been used during the Boer War to purify drinking water and prevent waterborne diseases.

Science Museum Group. Object number 1980-532

Fig. 122
Wounded men at a British field dressing station during the Boer War, 1899–1902.

Though many soldiers contracted a number of different diseases, fewer soldiers than ever before required significant wound care or medical treatment. This was due, in part, to the 'first field dressing', which was introduced for the war. Consisting of two sterile dressings in waterproof packaging, they were applied immediately after wounding. If this was done correctly, they prevented infection. The wounds of the Boer War were far easier to treat than in the past. Improvements to weapon technology saw the use of high-velocity, jacketed bullets. This meant that soft-metal bullets (often lead) were encased inside a hard metal jacket. Though these may sound more dangerous than their predecessors, they created clean wounds, resulting in less tissue damage and contamination. Wound infection was far less of a problem here than during wars fought in European climates, as the dry South African soil carried a very low bacterial load, resulting in fewer cases of gangrene and tetanus.

Shortly before the war, in 1895, Wilhelm Conrad Roentgen discovered X-rays, and their medical uses quickly became apparent. The British army took nine X-ray units over to South Africa, enabling them to identify shrapnel in the limbs of their wounded men. This was the first British military campaign to employ X-ray units (figs 123 and 124). The units not only improved the medical care on offer to British soldiers, but emphasised British scientific superiority over the Boers – something that became especially important after news of the embarrassing military campaign spread through Britain. Other grandiose demonstrations of British science were the first use of military balloons, depicted in artistic impressions in our collection (figs 125 and 126).

The British army medical services were criticised heavily in the aftermath of the conflict, resulting in a formal inquiry. Its final report concluded that though mistakes were made, 'in no campaign have the sick and wounded been so well looked after as they have been in the Boer War of 1900'.[8] By the end of the war, there were better structures in place in the British army medical services than ever before, and War Office reorganisations allowed doctors a greater say in military planning. In 1902 the Army Nursing Service became Queen Alexandra's Imperial Nursing Service, formed by a royal warrant

Figs 123 and 124
Major W. Scott-Moncrieff was shot five times while attempting to outflank the Boers at the Battle of Spion Kop, 1900. The X-ray (left) shows lead left in his hand after a bullet ricocheted off his rifle. The image of his leg (right) shows his fractured tibia and fibia.

Wellcome Collection

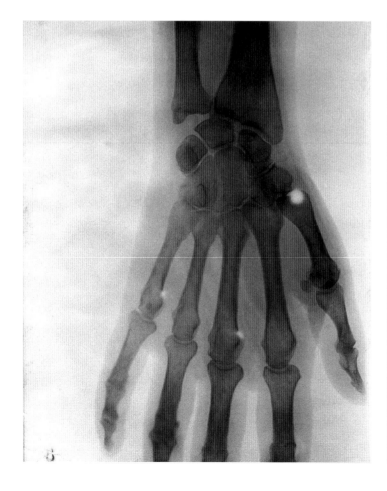

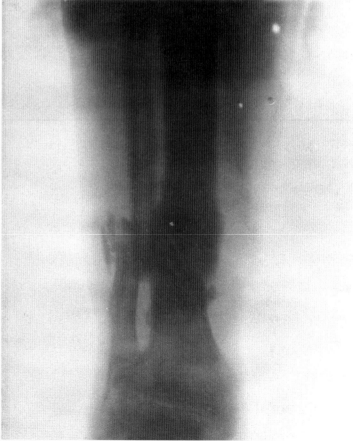

Fig. 125
Henry Charles Seppings
Wright, ink and wash on
paper, 1900. Drawn for the
Illustrated London News,
it depicts an experiment
to test the sustaining
power of military balloons.

Science Museum Group.
Object number 1979-525/3

OVERLEAF
Fig. 126
Frederick William Burton,
*Military Balloon during
the Boer War*, wash on
pasteboard, 1899–1900.
Drawn for the *Illustrated
London News*, it shows
cavalry in the field escorting
the observation balloon.

Science Museum Group.
Object number 1979-525/2

with the queen as their president. From 1906, hygiene and sanitation were taught to all officers and became a part of their examinations, and the Royal Army Medical Corps was reorganised to improve the efficiency of the service.

THE FIRST WORLD WAR (1914–18)

In 1914 the new and improved army medical service entered the First World War, confident of a quick and easy victory. The reality could not have been more different. The war lasted four long years and was fought all over the world. Over 70 million military personnel were involved from all participating countries, and more than nine million of these died, not including the seven million civilians who were killed as the lines between the civilian and military spaces blurred.

Though the experiences of the Boer War had convinced many of the importance of sanitation and hygiene, it lulled them into a false sense of security. The battlefields of the Boer War were completely different from the fertile fields of France and Belgium. Here, wounds festered and worsened, and many patients died from infection. To make matters worse, the wounds of the First World War were remarkably different from those encountered before. In the 100 years before the First World War weaponry changed from four-rounds-per-minute muskets, to 600-rounds-per-minute machine guns, resulting in patients with several severe bullet wounds, something never seen before. Others were hit by flying shrapnel, tiny bits of steel or lead encased in a hollow shell that exploded, propelling the shrapnel outwards at high speeds. Some were wounded by grenades and other artillery.

The development of poisonous gas, first used as a weapon in 1915, deeply affected soldiers. It burned their skin and irritated their lungs, prompting the creation of new gas masks for soldiers and animals alike (figs 127 and 128). Gas was just one of the highly effective new forms of weaponry used in the First World War. It was at this time that planes were first used as weapons; technology now allowed them to shoot bullets through their propellers and drop bombs on civilians and combatants alike. It was in this war, too, that the military tank was first introduced, inspiring the creation of terrifying protective masks worn by British tank operators (fig. 129).

When Allied soldiers were wounded on the Western Front, they were transported through the chain of evacuation. First they were tagged with their medical and military details (fig. 130). They then travelled through regimental aid posts, advanced dressing stations, field ambulances, casualty clearing stations and general hospitals. If they were severely wounded, they were sent to Britain to recover in hospitals and convalescent homes. Each stage was further away from the hostilities and provided more sophisticated care. At the beginning of the war, soldiers were moved between these points via horse-drawn ambulances. These were rickety and slow, and by the time they arrived at their destinations many patients had succumbed to their infections. The entirety of the British war effort was dependent on charitable people at home in Britain, and many of them donated their own cars to allow patients to be transported more efficiently between parts of the chain.

Doctors, nurses and untrained volunteers had to use their own judgement on each patient. They utilised a system called 'triage', where patients were divided into three distinct categories: those with minor wounds who could be patched up and made to fight again relatively quickly, those who could recover if given medical attention and those who would die regardless of medical intervention. Mary Borden, a nurse during the war, wrote:

> I was there to sort them out and tell how fast life was ebbing in them. Life was leaking away in all of them; but with some there was no hurry, with others it was a case of minutes.[9]

This was a difficult yet important decision. Though the Royal Army Medical Corps grew considerably during the war, doctors, surgeons and nurses were constantly busy. They had limited resources and were at the mercy of unreliable deliveries of equipment and drugs. It was vital that they did not waste time or medicine on men who were not going to survive.

Surgery was completed either at the Casualty Clearing Station or in a general hospital, depending on whether the patient was able to survive the wait. Many men had limbs amputated due to infection. Doctors hoped to save soldiers' lives by removing the infected limb before it killed them. In a world before antibiotics, practitioners did everything they could to prevent infection from

Fig. 127
Helmet-type gas mask, canvas, metal fittings and glass eyepieces, UK, 1915. This mask was used to protect wearers during a phosgene or chlorine gas attack. It was called the PH 'tube' gas helmet because of its appearance.
Science Museum Group. Object number A652157

Fig. 128
Gas mask for horses, Germany, 1914–18. Horses were used extensively during the First World War, and though gas was first used in 1915, it took a while for protective gear to be made for animals.
Science Museum Group. Object number A635089

Fig. 129
Protective mask, leather and chain mail, UK, 1914–18, worn by British tank crews.
Science Museum Group. Object number A204116

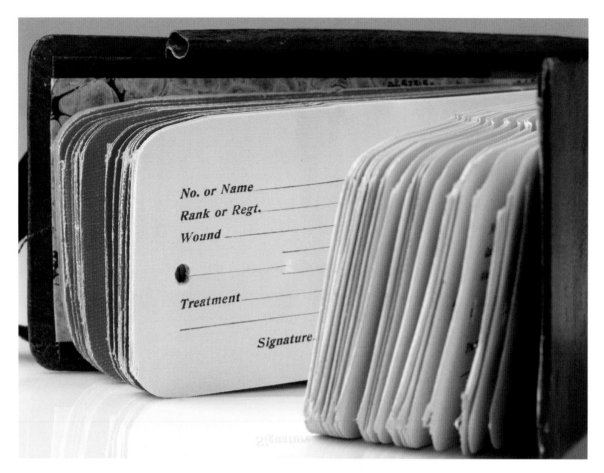

taking hold, but this was anything but easy. Wounds were washed with carbolic lotion, a mixture of carbolic acid and water, then wrapped in gauze and covered in more carbolic lotion. Doctors were often forced to use a variety of unconventional materials to wrap wounds, including petticoat cotton, curtain material, cotton muslin and sphagnum moss (the generic name for around 300 types of moss) (fig. 131). Many patients underwent the process of debridement, where dead or dying flesh surrounding the wound was cut away before the affected area was wrapped and allowed to recover. Though these methods were sometimes successful, it became apparent that there was a need for a better way of sterilising wounds.

The French surgeon Alexis Carrel (1873–1944) and the British biochemist Henry Dakin (1880–1952) worked on a new technique in 1915 and 1916. The Carrel-Dakin method was a form of wound irrigation. Dakin created a septic solution, while Carrel developed the apparatus used to apply it to the wound (fig. 132). The solution itself was notoriously difficult to get right. If the ratios were incorrect, it would not only irrigate the wound, but irritate it too. The famous pharmaceutical company Johnson & Johnson realised that hospitals lacked the resources, and crucially the time, to do this themselves. They mass produced ampoules of sodium hypochlorite and provided it to the British authorities, who officially adopted the method in 1917. The Carrel-Dakin method saved countless lives and limbs by preventing infection. In the long term, it has significantly changed the way we manage wounds. Prior to its invention, gas gangrene, a bacterial infection that causes gas to form in dying tissues, regularly resulted in either death or amputation.

Another key innovation of the First World War was the Thomas splint (fig. 133). It was invented by the surgeon Hugh Owen Thomas (1834–1891) in 1875, but it was not until 1916 that it was used on the fighting fronts. Thomas came from a family of Welsh bonesetters and had his own practice for treating fractures in Liverpool. Though it sounds like a relatively mundane medical object, the introduction of the Thomas splint lowered the death rate from broken femurs (thigh bones) from 80 per cent

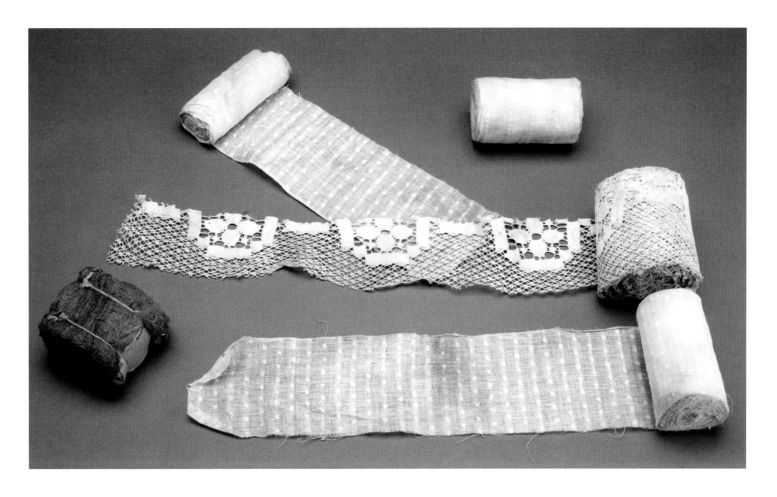

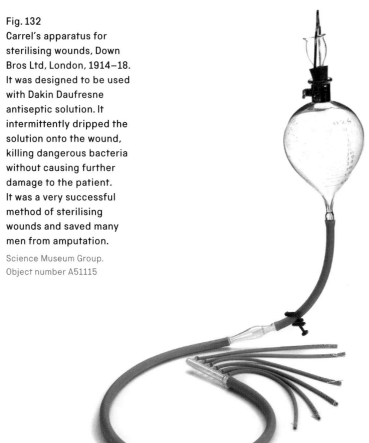

Fig. 132
Carrel's apparatus for sterilising wounds, Down Bros Ltd, London, 1914–18. It was designed to be used with Dakin Daufresne antiseptic solution. It intermittently dripped the solution onto the wound, killing dangerous bacteria without causing further damage to the patient. It was a very successful method of sterilising wounds and saved many men from amputation.

Science Museum Group. Object number A51115

to just 20 per cent. Prior to 1916 soldiers regularly died from blood loss from the femoral artery, because many of them went into shock, preventing medical practitioners from providing adequate care. The Thomas splint was easy to use and incredibly effective at immobilising the lower limb, thus allowing the patient to be moved without exacerbating the wound. The splint not only saved thousands of lives in the war, but it also reduced the number of amputations, thus exposing fewer soldiers to threat of infection.

One of the key problems of amputation and battlefield medicine more generally was blood loss. Though blood transfusions had been attempted for hundreds of years, often with animal blood and human patients, they failed because of coagulation, or clotting, which occurs when the donor blood is rejected by the recipient's immune system due to incompatible blood types. Austrian biologist Karl Landsteiner discovered three different blood types in 1901 (A, B and O), and learned that matching the blood group of a donor to that of the recipient prevented the patient from dying. Though doctors were aware of this during the First World War, blood transfusions remained dangerous. Without the ability to store blood properly, the donor and the recipient had to lie next to each other with their blood vessels connected via rubber tubing (fig. 134). This was revolutionised in 1914, when it was discovered that adding sodium citrate to blood prevented it from clotting, allowing it to be stored and used later. This was improved upon by the discovery of heparin in 1916. Both sodium citrate and heparin are anticoagulants, meaning that they block the creation of the proteins that cause clotting. Storing the blood with citrate and keeping it on ice allowed it to be kept for 26 days. This meant that the authorities were able to send blood to where it was most needed. While blood banks did not occur in civilian life until the 1930s, they became an integral part of military medicine during the First World War, mainly pioneered by Captain Oswald Hope Robertson, who established 'blood depots' in 1917, during his service in France. The ability to perform blood transfusions saved thousands of soldiers in the war, and the legacy of this conflict has given us the blood banks that continue to save lives every day.

Medicine was not just about saving lives or limbs, it also allowed literal and emotional scars to be concealed. The nature of the fighting in the First World War meant that more soldiers had facial wounds than ever before. Many were shot peering over the parapets of the trenches or were hit by shrapnel from artillery fire. The problem was so serious that the British army introduced steel helmets for the first time in 1916. Facial injuries not only created life-threatening wounds, but, if the patient survived, they were left with severe life-changing scars – stark reminders of a conflict most would rather forget. In cases of facial injury, surgeons close to the front line did what they could, but they did not have the time or skills to hide the marks left by their scalpel. Often the skin would heal and tighten, leaving men with permanently disfigured faces. A surgeon from New Zealand, Harold Gillies (1882–1960), opened a hospital in Sidcup, England, to reconstruct the faces of men that had been destroyed by war. He was inspired by an ancient Indian surgical technique that used the skin on the forehead to regrow the nose. His 'tubed pedicle graft' was an attempt not only to restore the health of his patients, but to make

Fig. 133
Leg splint, iron and leather, 1875–1920. Before the introduction to the war front of this 'Thomas splint', designed by Hugh Owen Thomas in 1875, nearly 80 per cent of men with upper-leg fractures died.

Science Museum Group. Object number A603026

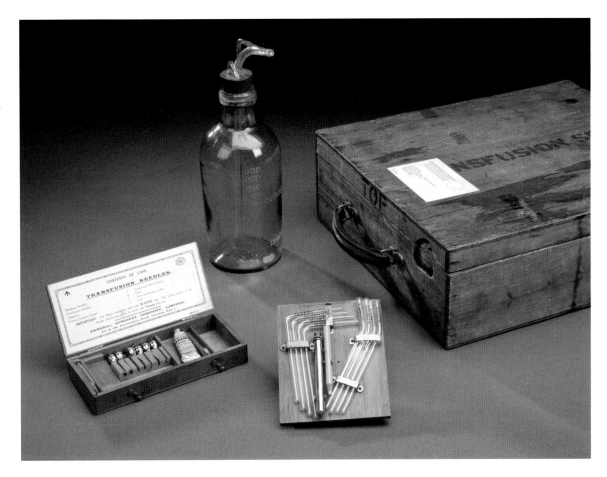

the results as aesthetically pleasing as possible.
This technique involved taking a flap of skin
from the chest or forehead, rolling it into a tube,
and attaching it to the area that required the
graft. This reduced infection rates by maintaining
the original blood supply. Gillies undertook his
work in stages, allowing the body time to heal
and recuperate between operations. His work
was so influential that he is often considered
the father of plastic surgery, and his work helped
to socially rehabilitate the soldiers whose scars
and wounds affected their emotional states.

It was not just surgeons who tried to help men
with the difficult process of social rehabilitation.
Francis Derwent Wood (1871–1926), originally
a sculptor, created tin masks for the facially
disfigured. These masks included a number
of different features, including eyes, eyebrows,
cheeks and noses. There are very few left in
existence, as many men became so accustomed
to wearing them that they were buried in them
when they died. Some soldiers utilised artificial
eyes after undergoing facial reconstructive
surgery (fig. 135). All of these cosmetic interventions
helped soldiers to recover emotionally from the

toil of war, allowing them to conceal their scars
if they preferred and avoid the social stigma of
disfigurement (fig. 136).

Medicine and surgery advanced significantly
in the four years it took for the war to come to
an end, and a number of these developments
are still relevant to medicine today. It is due to
the First World War that we have civilian blood
banks to draw from, and that we know more
about the importance of antiseptic wound
treatment in military situations and how
conditions on the battlefield significantly
impact the health of the army. One of the key
outcomes of the First World War was a better
understanding of physical rehabilitation and
the usefulness of occupational therapy, as well
as mental trauma, a topic too large and nuanced
to study in this chapter. The sudden influx of
thousands of disabled British soldiers forced
the government to provide some basic help.
A lucky few were able to draw upon basic
state pensions, while some were provided with
prosthetic limbs and given occupational therapy
to prepare them to return to the workplace
(see chapter 2).

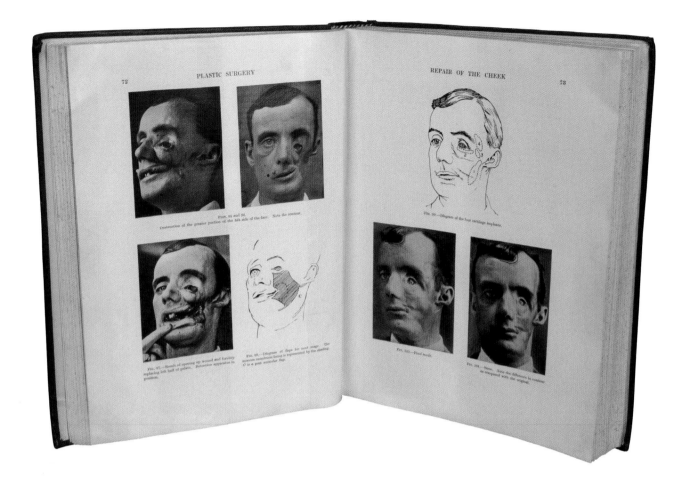

Fig. 136
Harold Gillies, *Plastic Surgery of the Face*, 1920. Compiled by Gillies, a pioneer in facial reconstruction, this training manual was published after the First World War for surgeons wishing to specialise in plastic surgery.

THE SECOND WORLD WAR (1939–45)

The Second World War was the deadliest military conflict in global history. While the First World War ushered in increased targeting of civilian populations, during the Second World War this was experienced on an unprecedented scale. Though the death toll is uncertain, it is estimated that between 50 million and 80 million people died, including civilians. Far fewer British military personnel were wounded in the Second World War than in the First, but the public experience of this conflict provided the impetus for medical innovation.

One of the most significant medical developments during the war was the mass production of penicillin. This antibiotic was used extensively in the conflict to treat venereal diseases (sexually transmitted infections, known commonly as VD) and also for infected wounds. VD caused a significant waste of military power during the Second World War, especially for soldiers stationed abroad, who were allowed (if not encouraged) to visit *maisons de tolerance*, a term used to describe the brothels that were tolerated by the authorities. Methods of treatment varied, but most relied on the use of sulphonamides, which were synthetic anti-microbial drugs that prevented some types of bacteria from multiplying. The first was called Prontosil and was initially used in experiments in 1932. Though it was effective, recovery times were slow, robbing the army of much-needed manpower. Penicillin was first discovered by Alexander Fleming in 1928, but it took until the early 1940s for a process of mass production to be perfected. The Science Museum collection has a sample of penicillin mould that Fleming gave to Douglas Macleod a few years after discovering its antibiotic potential (fig. 137). Penicillin had a significant impact on the treatment of wounds and disease in the Second World War. It meant that infection posed fewer problems than ever before, and VD was treated far more quickly, allowing men to return to battle. Prior to the use of penicillin, it had taken weeks of medical care and hospitalisation before a soldier was fit to fight again. The Allied forces used penicillin wherever they could: Major Thomas Scott's penicillin chest travelled all the way to North

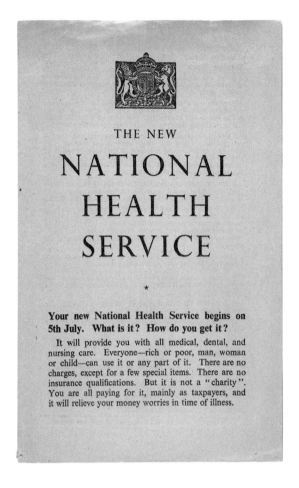

Fig. 137
Sample of Penicillium mould presented by Alexander Fleming to Douglas Macleod in 1935.

Science Museum Group.
Object number 1997-731

Fig. 138
Wooden chest, 1942–5. Owned by Major Scott Thompson of the Royal Army Corps, this was used to transport penicillin supplies to North Africa during the Second World War.

Science Museum Group.
Object number 2013-64

Fig. 139
National Health Service leaflet. Delivered to every home in Britain in 1948, it detailed the formation of publicly funded free healthcare to which everyone was entitled.

Science Museum Group.
Object number 2018-477

Africa (fig. 138). The significance of penicillin is hard to overstate. By the end of the 1940s more than 250,000 people were prescribed it every month. It made previously untreatable conditions curable and allowed surgeons to attempt more invasive surgeries than ever before. Since the discovery of penicillin, medical science has identified over 100 different types of antibiotics. Worryingly, we are now facing antibiotic-resistant bacteria because of over-use. Some forms of bacteria have grown accustomed to the antibiotics used to treat them, threatening to render our main method of fighting infection obsolete. There is a real risk that antibiotics will one day, perhaps soon, no longer be effective, and we will again be presented with the challenges faced by the pre-antibiotic world.

Penicillin was not the only important discovery of the Second World War. The surgeon Archibald McIndoe worked on new treatments for patients who had suffered severe burns. McIndoe trained with his cousin Harold Gillies and later established a hospital in East Grinstead, West Sussex, for pilots who had crashed their planes. There, he rebuilt 'faces, building new noses,

eyelids, chins and cheeks and turning the useless remains of burned fingers into usable stumps'.[10] But he is more often celebrated for the social rehabilitation that his work offered patients. He removed all the mirrors in his hospital to avoid distressing patients by their appearance and encouraged them all to explore the local area, having instructed the locals not to gawp or stare. McIndoe's patients referred to themselves as the 'Guinea Pig Club', reflecting the experimental nature of his work, much of which built upon the foundations of modern plastic surgery, laid by his cousin Gillies, 30 years before.

Though there were many additional pressures for social reform, the Second World War was, to some extent, responsible for the formation of the National Health Service by the Labour government in 1948. The experience of war meant that the British people were accustomed to state intervention in their lives, and the number of civilian casualties pressurised the government into acknowledging their responsibility to keep citizens safe and healthy, while the introduction of rationing inadvertently improved the health of the poorest members of society. Three years after the end of the war, civilians received leaflets in the post informing them that they were entitled to free healthcare, changing the political and medical landscape of Britain forever. Though the NHS continues to change, it remains an integral part of Britain's culture and international reputation (fig. 139).

MODERN MILITARY MEDICINE

Though medicine improves and evolves quickly, weaponry and warfare changes even quicker, necessitating further innovation in medical care. It is worth noting that several modern-day innovations in military medical equipment are adaptations of and improvements on existing historic objects.

To this day, blood loss is the most common cause of death on the battlefield. It is incredibly important for bleeding to be controlled as fast as possible. The C-A-T tourniquet is a good example of a piece of modern equipment that has been developed from an older idea (fig. 140). Though tourniquets have been used for thousands of years, experts have worked tirelessly to improve the efficiency of the design. The C-A-T tourniquet has reduced the rate of mortality from blood loss in the US army by 85 per cent since its introduction in 2005. It was designed to be used one-handed so that patients can apply it to themselves, even when under fire. If the pressure applied by the tourniquet is insufficient, haemostatic products, like Celox, are used at the same time. These promote clotting to prevent blood loss. Celox is used in the British army and more recently in civilian medicine. It contains chitosan, a sugar extracted from the outer shells of shellfish and mixed with an alkali. When it encounters blood, it gels to create a barrier, thereby preventing bleeding (fig. 141). The so-called 'Prometheus' pelvic splint (fig. 142) is another brand that has been adapted from existing technologies. Prometheus has built on the legacy of the Thomas splint to improve its efficacy and the ease with which it can be used. Scientists, doctors and engineers are constantly improving the medicine and resources available on the battlefields. Key areas of research include tissue regeneration, improved sterilisation, infection

Fig. 140 (below left) Combat Application Tourniquet (C-A-T), 2005–17. Building on older technologies, this one-handed tourniquet can be used on a traumatic wound to prevent extreme blood loss.

Science Museum Group. Object number 2018-86

Fig. 141 (below right) Celox products, MedTrade Products, UK, 2016. These newly developed wound dressings and haemostats can help to control bleeding from arterial injuries.

Science Museum Group. Object numbers E2016.0406

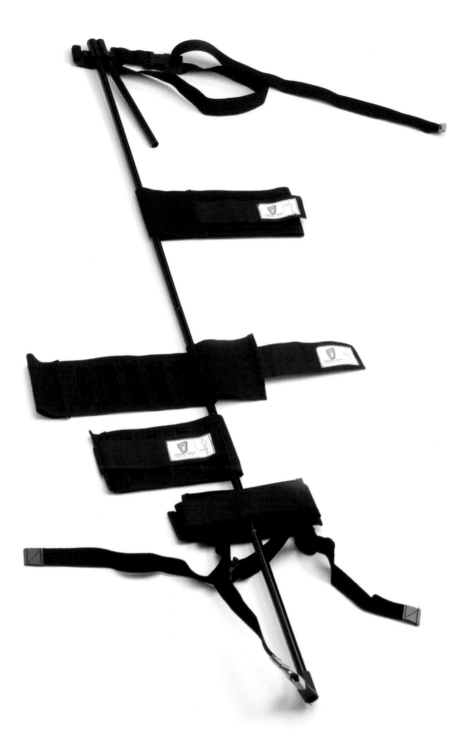

control, assistive technology (including prosthetics) and both mental and physical trauma treatments.

While the Science Museum's collections are stocked with medical objects from many conflicts, it is a significant challenge to curatorial practice to acquire cutting-edge military equipment. These items are kept secret and confidential before being introduced. They are also constantly changed and updated to ensure that they are as successful and efficient as possible. Thus, curators are reliant on the relationships we build with scientists, engineers and academics: we visit labs and attend conferences, or we wait until the objects become available in the civilian medical market, like the Celox and C-A-T products. It is one of the great challenges – and joys – of curatorial work: it is never finished, and contemporary collecting goes on.

Each of the medical advances discussed in this chapter is a strong testimony to human ingenuity in the face of war. Conflicts provide new medical problems that require fresh and innovative thinking. Military medicine progresses at a rapid rate – as it must. From humble beginnings, it has now become a highly respected sub-section of the medical profession.

Fig. 142
Pelvic splint, Prometheus Medical, UK, 2015. The splint helps immobilisation and therefore reduces the risk of haemorrhage. It is used widely in emergency medicine.

Science Museum Group.
Object number E2015.0489.3

6
PREVENTING CERVICAL CANCER
VAGINAS, VACCINATION AND VENEREAL DISEASE

EMMA STIRLING-MIDDLETON

In 1964 the British government established a national programme that made available a free, five-minute test designed to prevent women from developing cervical cancer. It is estimated that, if implemented successfully, cervical screening should prevent around 83 per cent of deaths from cervical cancer. Five decades later, at the time of writing, nine women in the UK are diagnosed with cervical cancer every day and 900 die from it each year. Rates of cervical cancer for young women have risen dramatically over the past 20 years, making it the most common cancer among women under the age of 35.[1] For half a century, the UK population has had the means to make cervical cancer a thing of a past – so why is the number of women experiencing it increasing?

The answer is that many women are simply not turning up for the test. To understand why women are making this choice, we must delve into a 200-year story of our developing understanding of cervical cancer, changing screening programmes and the social contexts in which they emerged. Ultimately, this exploration will demonstrate that diseases, and our responses to them, are deeply social phenomena.

'THE LOWEST DEEP INFAMY AND DEGRADATION': EXAMINING THE VAGINA IN 19th-CENTURY BRITAIN

From an early age, society teaches us that medical knowledge is an objective, unquestionable truth, an immutable presence awaiting human discovery. This is a dangerous myth. Medical knowledge is produced by people in the same way that art, politics and music are. This is not to say that there is not truth in medicine; on the contrary, it is a cornerstone of human life that enables us to survive, thrive and better understand ourselves and the world around us. The point is that we must recognise that medical knowledge is constructed by people. From invention, manufacture and dissemination to interpretations, uses and impacts, medical ideas shape and reflect the societies within which they exist. All ideas achieve significance in relation to other ideas. By exploring some of the key medical technologies encountered by women undergoing tests for cervical cancer over the past century and bringing into focus the social contexts in which they were created and imbued with meaning, we can better understand attitudes towards cervical screening today.[2]

The oldest and simplest gynaecological tool is one that we still use: the speculum. A speculum is a medical tool used to open a woman's vagina for inspection. Although it has been used since antiquity, with examples dating from around two thousand years ago found among the remains of the Roman city of Pompeii (fig. 143), the speculum largely fell out of use during the medieval period. It was not until the 19th century that the speculum made its comeback. In 1801 Joseph Récamier (1774–1852), Professor of Medicine at the Collège de France in Paris, developed a smooth, reflective tube made from tin to help him examine a patient with an unusual vaginal discharge (fig. 144).[3] The instrument, designed to help doctors to see the vagina and cervix, marked a significant break from traditional genital examination, which was generally a hurried and awkward affair conducted using touch and not sight in an attempt to protect a woman's 'modesty' (fig. 145). Many people, both experts and laypeople, initially rejected Récamier's speculum as inappropriate and morally degrading (fig. 146). A British pamphlet of 1857 advised husbands that the use of a speculum to examine a woman's vagina 'plunges its wretched victim ... down into the lowest deep infamy and degradation'.

The early 19th century saw a significant increase in incidences of syphilis and gonorrhoea, prompting a surge in public anxiety around sexuality (figs 147 and 148). Besides the fact that they seemed to be transmitted through sexual activity, the medical establishment knew very little about these diseases and were thus unable to take action to prevent their spread. At the time, sexually transmitted diseases were known as 'venereal diseases'. This term originated from the name of the Roman Goddess of love and desire, Venus, and clearly associated sexually transmitted disease with females. An epidemic of sexually transmitted disease posed a grave threat to Victorian notions of gender and sexuality. Society was structured around the woman as a fragile 'angel of the house' in need of care and protection, and the man as healthy, virile and active, provider and protector. Female sexuality was upheld as marital duty, performed without pleasure for the purposes of reproduction and the satisfaction of male desire. How could women remain pure and benign without framing men as the locus of responsibility for sexually transmissible diseases? The body of the prostitute was the answer. Sexually transmitted disease was attributed to a fictional stereotype of the depraved and promiscuous prostitute. Stigmatising prostitutes as the source of contagion framed 'ordinary' men and women simply as victims of a social evil. In the absence of a means to control disease, the state sought to control the body of the prostitute, with the speculum its primary tool.

[The prostitute] is a woman with half the woman gone, and that half containing all that elevates her nature, leaving her a mere instrument of impurity ... a social pest, carrying contamination and foulness to every quarter to which she has access.

William Acton, *Prostitution, Considered in its Moral, Social and Sanitary Aspects*, 1857

In France, prostitutes were officially regulated by the authorities. From 1810, they were required by law to register with the police and submit themselves for vaginal examination with a speculum every two weeks. If a prostitute was found to be diseased, she would be detained and treated in a prison hospital. Failure to comply resulted in incarceration.[4] This system became well known throughout Europe and soon spread to Britain, where the Contagious Diseases Acts (1864, 1866 and 1869) similarly required prostitutes to submit to internal examination with a speculum.[5] The Acts focused on military garrisons and naval towns and entitled policemen to hospitalise any woman they suspected might be a prostitute with a sexually transmitted

Fig. 144 (below)
Vaginal speculum with mirrored interior of the type developed by Récamier. S. Maw, London, 1851–1900.

Science Museum Group. Object number A646883

Fig. 145 (opposite, left)
Vaginal examination, print by J. P. Maygrier, 1825. A doctor averts his gaze while examining his patient.

Wellcome Collection

Fig. 146 (opposite, right)
Félicien-Joseph-Victor Rops, Doctor examining a woman's vagina using a speculum, pen and ink on paper, 1850–98.

Wellcome Collection

Fig. 147 (opposite, below)
Four wax anatomical heads showing the effects of syphilis on the male body, Germany, 1910–20.

Science Museum Group. Object number E2001.578.4

Histoire générale du toucher

PL. XXIX

Toucher, la femme debout

Speculum (E. 759)

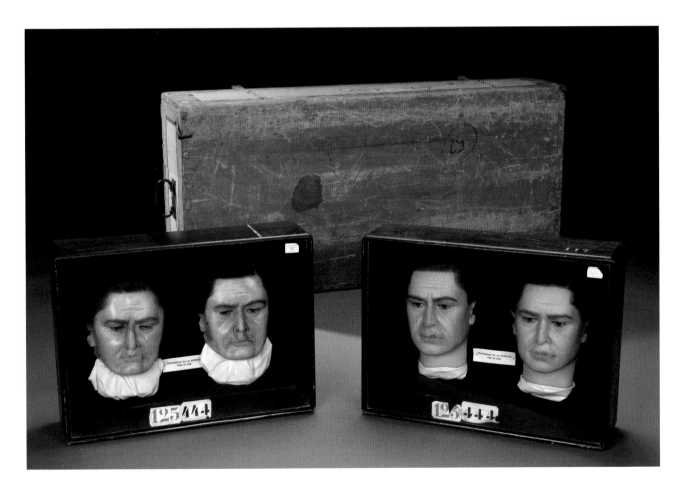

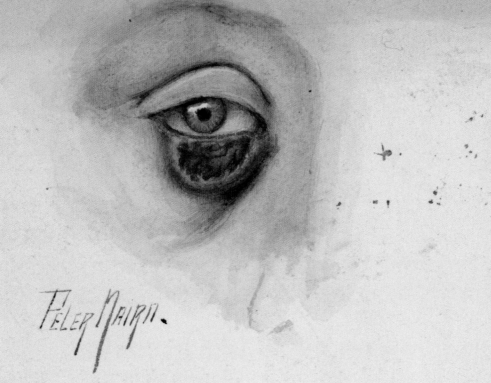

Peter Nairn.

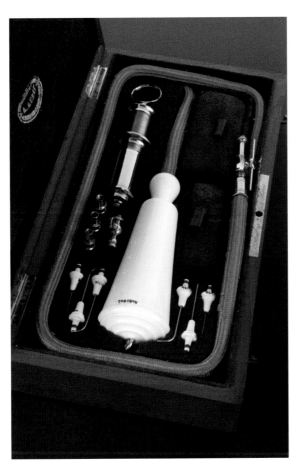

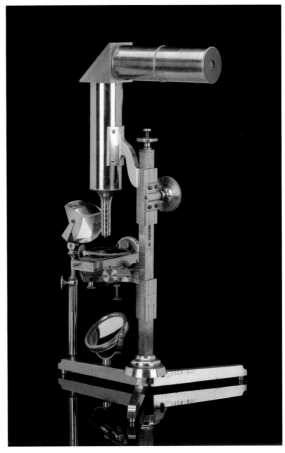

Fig. 148 (opposite)
Peter Nairn, A syphilitic
sore, watercolour,
1891–1920.

Science Museum Group.
Object number A680449

Fig. 149 (right)
Mercury douche, France,
1840–95. The douche
was used to inject mercury
through the urethra to
treat venereal disease.

Science Museum Group.
Object number A680862

Fig. 150 (far right)
Microscope used by the
founder of cell theory,
Theodore Schwann, 1835.

Science Museum Group.
Object number 1928-801

infection. Diseased prostitutes were sequestered in hospital for up to nine months to prevent the spread of infection.[6] The maximum term was nine months to allow time for any pregnancies to come to term. Mandatory treatments were given, such as the administration of high doses of mercury, both internally and externally, and the application of substances to burn off sores or growths (fig. 149). We now know that mercury is highly toxic and can cause damage to the nervous system, brain and other major organs. Women could also be submitted to moral and religious instruction while detained in hospital. On release, they could be forced to attend medical examinations every two weeks for up to a year.[7] The new laws regulating prostitution were perceived by many as a positive mechanism for the control of sexually transmitted disease.

Récamier's speculum significantly improved the efficacy of clinical observation of the female reproductive organs. However, from the outset, the instrument became synonymous with promiscuous sexuality and sexually transmitted disease. At the time, cancer of the cervix had not been identified as a distinct disease and women

were instead diagnosed with 'cancer of the womb', a general term for cancers of the female reproductive organs. Récamier advocated the use of the vaginal speculum to apply fruit extracts and opium-based concoctions to cancerous ulcers of the vagina and uterus, drawing cancers of the womb into the sphere of sexuality and morality in the medical imagination.[8] In the space left by a lack of understanding of what cancer of the womb was, why it developed and how it could be treated, doctors fell to speculation that was strongly influenced by public discourse. Some doctors explicitly connected cancer of the womb to sexual deviance, such as the Canadian doctor Guillaume Vallée, who, in 1826, noted that 'Lower class women who live in cities are decidedly more affected by [cancer of the womb] than those who live in the countryside… and how can one explain such a difference if not by their greater moral laxity?'[9] Accusations such as these held women responsible for their development of cancers of the reproductive system. As with syphilis and gonorrhoea, the medical establishment sought control over a distressing affliction by framing a certain section of society as responsible for, and therefore susceptible to, the disease, demarcating those at risk from developing cancer of the womb from the majority. This stigmatisation bolstered the authority of the medical establishment, creating the illusion of solutions where there were none. It also acted as an instrument of social control, using the fear of disease to police women's bodies and behaviour.

CERVICAL CANCER GETS CELLULAR

Although microscopes were first developed in the 1600s, it was not until the 19th century that they became reliable and powerful, opening up the cellular world to scientific investigation (fig. 150). Doctors had begun to examine human tissue under the microscope and were recognising that the cells that made up cancerous tumours looked different from those found in healthy tissues. Early 20th-century doctors such as Carl Ruge (1846–1926) and Johann Veit (1852–1917) began to advocate for 'microscopic diagnosis of cancers of the womb'.[10] They advised doctors to use a speculum to examine the cervix and take a biopsy of any lesions they identified, for examination under the microscope. The ability to see the microscopic world and identify the difference between 'normal' and 'abnormal' cells introduced a new era of diagnosis.

Microscopic analysis of cells from different areas of the reproductive system facilitated more precise diagnoses: in other words, doctors could identify that a woman had healthy cells within her cervix while having abnormal cells within her uterus, and vice versa. For the first time, doctors could move away from the generic 'cancer of the womb' and differentiate between cancers of the cervix and the uterus. A growing scientific understanding of gynaecological cancers began to shift the focus of medical discourse and treatment away from female behaviour and onto a new frontier: the cell.

THE 'PAP SMEAR' TEST

[T]he present difficulty in accomplishing early diagnosis lies in the fact that we must depend on subjective symptoms of the disease to bring the patient to the physician, and by the time the patient becomes sufficiently aware of discomfort to seek help, the disease is far advanced ... if by any chance a simple, inexpensive method of diagnosis could be developed which could be applied to large numbers of women in the cancer-bearing period of life, we would be in a position to discover the disease in its incipiency much more frequently than is now possible.

George N. Papanicolaou and Herbert F. Traut, 'The Diagnostic Value of Vaginal Smears in Carcinoma of the Uterus', 1941

The story of cervical cancer screening begins in Greece, aboard a ferry to Athens in 1910, when a man named George met a woman named Andromache (known as 'Mary'). George Papanicolaou (1883–1962) was a violinist and a young doctor who had recently received his PhD in zoology, and Mary Mavroyeni (1890–1982) was a talented pianist from a high-ranking military family. The young couple fell in love and were married that September. On 19 October 1913, without any plans, unable to speak English and armed with just $250 between them, George and Mary moved to New York. They made ends meet by doing odd jobs: Mary sewed buttons at a department store and George sold rugs, and played the violin in local restaurants. George managed eventually to get a job as an assistant at the pathology and bacteriology department of the New York Hospital. By 1914, both Mary and George were working at Cornell University Medical College's Anatomy Department, George as a research assistant and Mary as his technician. The University would not permit a husband and wife to work in the same department and so Mary offered her services as a volunteer. Mary's work was never formally acknowledged.

George completed studies on guinea pigs as part of his research on the reproductive cycle. Each day he extracted cells from the vagina of a guinea pig, smeared them onto a glass plate and examined them under a microscope. He discovered that by looking at the changes in the guinea pig's cervical cells over time, he could predict successfully when the guinea pig would ovulate. Interested to know if the technique could be applied to people, George began to study the cervical cells of a woman that he referred to as 'a special case'. It was later revealed that this 'special case' was in fact George's wife, Mary.[11] He collected cells from her cervix (the entrance to the womb from the vagina), 'smeared' them onto a slide, and examined them under a microscope. Mary became the first person to undergo what would later be known as the 'Pap smear' test (referencing George's surname, Papanicolaou, and the smearing motion needed to transfer cervical cells onto a glass slide).[12]

Encouraged by the results, George began to use his method to systematically examine more human vaginal cells. In February 1925, he received funding for a project on the study of vaginal cells. Twelve women, mostly female workers at the New York Woman's Hospital, volunteered to undergo a daily smear test for two to three months. George examined their cells every day to understand the ways in which cervical cells can change over time (fig. 151). Mary was also skilled at examining cells under the microscope and they worked together on this research. Once they had achieved a thorough understanding of healthy cervical cells and the ways that they can change over time (fig. 152), they began to study the cervical cells of women with gynaecological cancers. Through this extensive study, they were able to determine the difference between normal and abnormal cells. Using this knowledge, George developed a test that used microscope observation of cells from a woman's cervix to diagnose cervical cancer. Vitally, George identified that cervical cells begin to change before cancer has developed. These 'pre-cancerous' cells had been entirely undetectable previously as they produce no tangible external symptoms. This discovery meant that these cells could now be treated, preventing them from becoming cancerous.

George presented these initial findings at the Third Race Betterment Conference in Battle Creek, Michigan in 1928.[13] Perhaps surprisingly, this conference was a gathering of eugenicists, a group who sought to improve the genetic stock of the human race by controlling reproduction. It is likely that George chose to reveal his research at this particular event because at the time, he was working under Charles R. Stockard, the chairman of the Cornell Medical School department of anatomy and a researcher in experimental genetics. Indeed, it was within

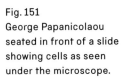

Fig. 151
George Papanicolaou seated in front of a slide showing cells as seen under the microscope.

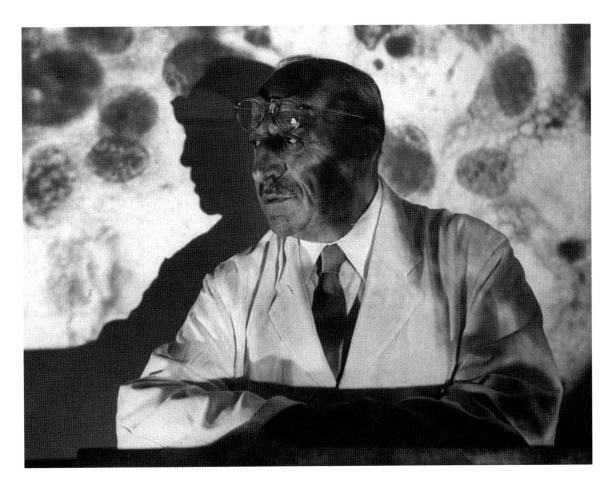

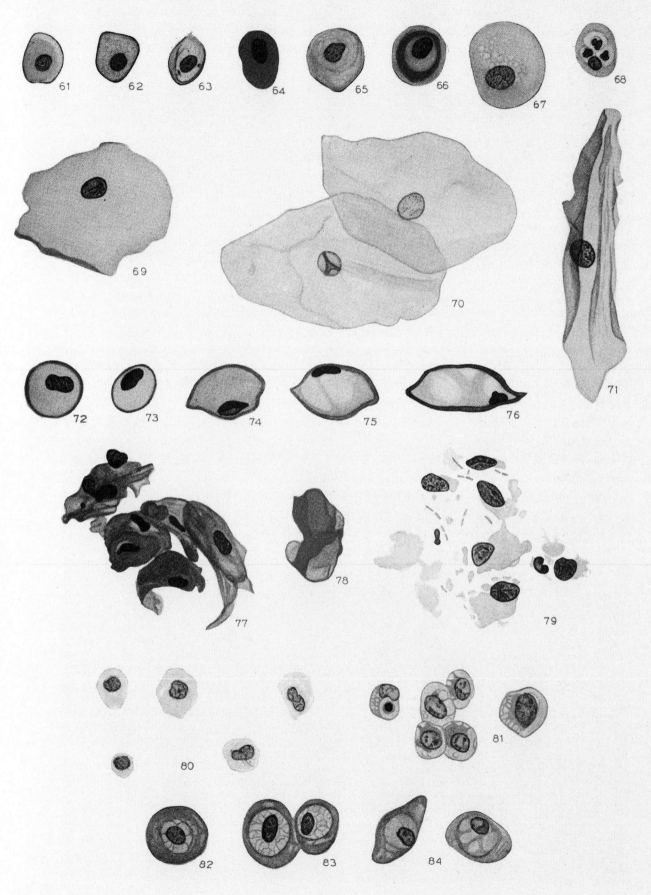

Fig. 152
Drawings of normal
vaginal smears by George
Papanicolaou, 1933.

Wellcome Collection

this department that George had completed his earlier research on guinea pigs. Such encounters remind us that medical research can be employed and interpreted in a multitude of ways: the line between the 'right' and 'wrong' sides of history can be so very thin. Following further research, George collaborated with the American gynaecologist Herbert F. Traut to publish further evidence of the value of his test for diagnosing cervical cancer in the *American Journal of Obstetrics and Gynecology* in 1941.[14]

George examined Mary's cervical cells on a regular basis over a period of more than 20 years. Her cervical cells remained normal and regular throughout her life and therefore enabled George to map the lifecycle of healthy cervical cells. He published his seminal work, *Atlas of Exfoliative Cytology*, in 1954. George received 18 nominations over four years (1949–53) for the Nobel Prize in Physiology or Medicine, though he never did win. Mary, dubbed by the gynaecologist George A. Vilos as 'undoubtedly ... the most over-tested woman of all time',[15] was presented with an award for being a 'Companion to Greatness' by the American Cancer Society in

1969. One might suggest this was rather too little, too late. Mary Papanicolaou's silent and unrecognised contribution to our understanding of cervical cancer is discomfiting to modern eyes. While the lack of written records from Mary's perspective makes it difficult for us to determine the full extent of her work, a reconstruction of her career and point of view from the available sources is a long overdue avenue of historical research. Issues of due credit aside, the Papanicolaou test changed the world, saving countless women from developing cervical cancer.[16]

By the year of George Papanicolaou's death in 1962, cervical cancer was a major cause of death among women in Britain. The Pap smear test was inspiring hope and rallying calls within the medical community. In the same year, Dr Marie Grant, a Scottish medical officer of health who had worked in an early centre for cell screening that had opened in Edinburgh in 1949, stated that 'cancer of the cervix must now be regarded largely as a preventable disease, and, with the remedy in our own hands, surely few can fail to be impressed with the urgency of the problems presented to us'.[17]

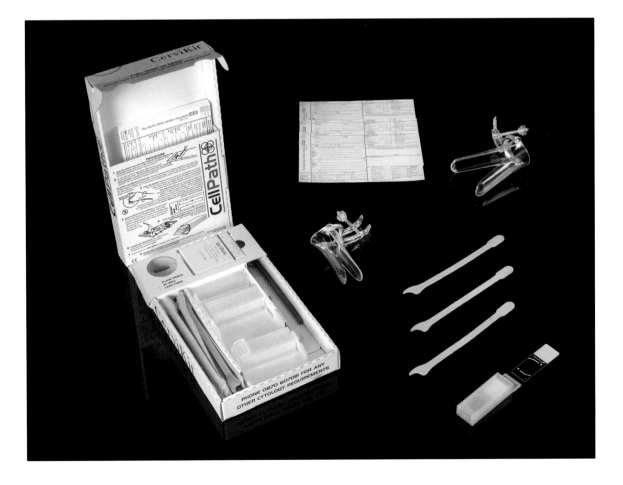

Hugh McLaren, Professor of Obstetrics and Gynaecology at the University of Birmingham, took this further, addressing colleagues at a conference in 1964 with his 'epitaph for cervical cancer':

we now have a method of prophylaxis which, if properly used, may largely eliminate [cancer] of the cervix from our communities. We can now say to womankind, 'If you submit to regular tests you won't get cancer of the cervix.'[18]

In 1948 the National Health Service had nationalised the hospital service in Britain with the aim of providing good healthcare for all, regardless of wealth. The service was governed by three core principles: 'that it meet the needs of everyone, that it be free at the point of delivery [and] that it be based on clinical need, not ability to pay.'[19] This medical system, in its infancy at the time of Papanicolau's publications, provided an effective national framework through which to introduce the Pap smear test. In 1964, almost 40 years after Papanicolau first presented his findings in the United States, the National Cervical Cytology Screening Service was established in Britain for self-referring women aged over 35 (fig. 153).

The test involved widening the vagina with a speculum, then using a wooden spatula to gently scrape away cells from the cervix. The cells were then transferred to a glass slide for examination under a microscope at a laboratory (fig. 154). Technicians examined manually every cell on each woman's slide (around 80,000 cells per smear test), looking for abnormal cells that could be a sign of cervical cancer. This was a time-consuming and laborious process, which required a large, skilled workforce. With intense focus, a technician could examine one smear in roughly five minutes, and around ten smears within an hour. Married women who stayed at home while their husbands went out to work were seen as a potential pool of labour that could be mobilised at short notice, and this demographic was targeted directly in calls for laboratory technicians.

The National Cervical Cancer Prevention Campaign noted in its 1966 newsletter:

It has been proposed that married women technicians, not working for domestic reasons, could be called upon to do part-time screening in local laboratories and a campaign is being launched to encourage them to come forward.[20]

Marilyn Symonds, who worked at the cytology laboratory at the Stoke Mandeville Hospital in 1964, described the workforce at the time as:

mainly married women with children, who wanted an interesting job to fit around their home commitments. A lot of them had had some scientific background of working within perhaps research, before they had their children, or perhaps had worked in school laboratories.[21]

The sociologists Monica J. Casper and Adele E. Clarke are critical of this feminisation of the role of technician, pointing out that laboratories targeted women specifically because they could be paid a lower wage than men, despite the fact that the job of technician was highly skilled and had the highest of stakes, with mistakes potentially meaning the difference between life and death. The exploitative nature of this recruitment drive led them to dub cervical screening laboratories in this period as 'technological sweatshops of late capitalism'.[22]

While the introduction of a national screening programme was a significant step towards reducing cervical cancer mortality rates, in reality the programme was inconsistent, inefficient and uncoordinated, and cervical cancer was not 'largely eliminated from our communities', as many had hoped it would be.

In 2008 the Wellcome Trust held an oral history Witness Seminar, which explored the history of cervical cancer from 1960 to 2000. The seminar brought together 28 participants who had played a part in the history of cervical cancer at some point in their lives, including clinicians, scientists, technicians, historians, statisticians and others interested in the recent history of cervical cancer. Many of the participants described their experiences of the early days of the National Cervical Cytology Screening Service. For example, the academic pathologist Professor David Jenkins described the haphazard organisation of the new national service: 'One of the things that intrigues me about the very early days of cytology is that it really wasn't organized in any national sense, or in any sense above that which was generated by the consultants themselves.'[23] Reports suggest that the training offered to those responsible for delivering the service was inadequate and varied dramatically from region to region. Professor Dulcie Coleman, who began her training in clinical cytopathology in 1962, described her experience of preparing to work for the National Cervical Cytology Screening Service:

> There were really no formal plans of [sic] how this should be done, but every pathologist became responsible for this. I think there were four or five training schools set up ... It was fairly basic training, because you were shown slides, and told: 'This is abnormal and this isn't.' There was no one to explain to you what you were looking at or to explain that the aim of the exercise was to detect a precancerous condition.[24]

Vitally, there was no centralised national policy or standards for cervical screening and consequently there was no consensus across the country about who should be screened, how often and how to classify the results. Professor Albert Singer, Professor of Gynaecology (Emeritus) at University College London, demonstrated this at the Witness Seminar, noting:

> One day [Stanley Way] told me this story of how it came to be that there was a five-year screening interval – the Government used to say it was five years – and he said, 'Well, they asked me one day down to a meeting and they said how many years do you think there should be for an interval?' and he said: 'Well, I really didn't know, there was no evidence, so I thought five-yearly was a reasonable target.'[25]

Due to the lack of national policy, women were mostly tested opportunistically if they happened to be visiting the doctor for reproductive or gynaecological care already, and those who were tested rarely experienced adequate follow-up after the test.

Fig. 154
Cervical smear screening apparatus, 1960–65.

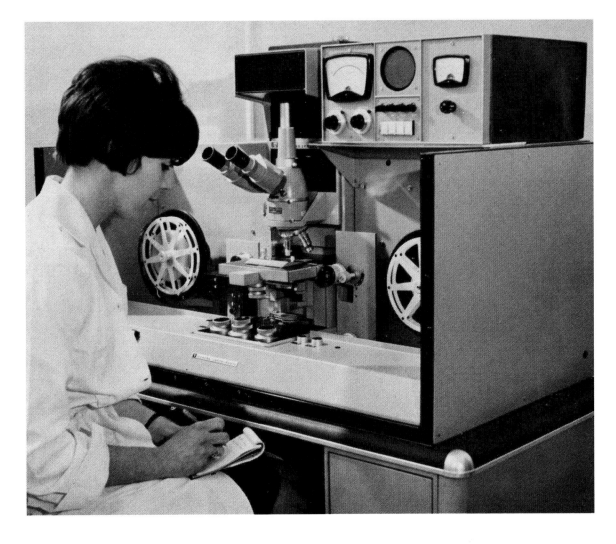

By the mid-1980s, it was clear that the government urgently needed to take action to improve the cervical screening programme. The number of deaths per year from cervical cancer had fallen only from 2,400 in 1965 to 2,100 in 1980[26] and, of greater concern, cervical cancer death rates for people aged 25 to 34 had actually trebled since the introduction of the National Cervical Cytology Screening Service in 1964. At least two-thirds of women diagnosed with invasive cervical cancer (cancer which has developed the ability to spread to other parts of the body) had never even been screened. The government's approach to cervical cancer prevention was not working.

In 1985 David Slater, a member of staff at the Department of Histopathology at Rotherham District General Hospital, published a scathing attack on the situation in the *Lancet*. His letter to the editor, entitled 'Cancer of the Cervix: Death by Incompetence', complained that cytopathology laboratories were underfunded and chronically understaffed and, as a result, were not operating effectively. He pointed out that the thousands of laboratory staff involved in examining the specimens needed to become

completely reliable if the screening programme was ever going to be worthwhile, and he called for a standardised call and recall system to be introduced to invite women for testing and follow up appointments.

In 1988 the government finally launched a centrally organised NHS cervical screening programme for every British resident who met certain criteria. Crucially, the programme no longer relied on women referring themselves to be tested. Women aged 20 to 64 were systematically invited for a Pap smear test every three to five years, and their attendance was managed by a computerised call and recall system that tracked when women needed to be invited for screening, their attendance record and details of any further testing or treatment that was required. Screening was co-ordinated at a local level in compliance with a centralised national policy. Clear guidelines were developed, providing a national standard for who should be screened, how and when, with a new focus on quality and consistency in both the process and the results. This egalitarian system did not discriminate according to race, disability,

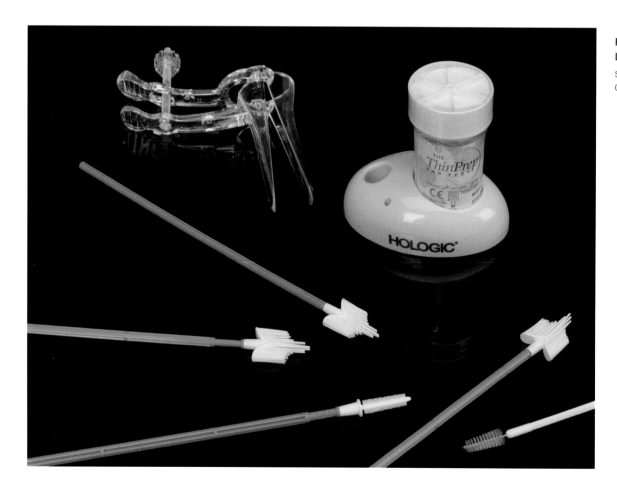

Fig. 155
Liquid-based cytology kit.

Science Museum Group.
Object number E2017.0169.16

occupation, sexuality or socio-economic background, communicating to the public that any woman within the eligibility age range could develop cervical cancer. By 1999, 80 per cent of women were being tested regularly and deaths from cervical cancer in Britain had fallen by 45 per cent – a striking success.

In 2003 a new Pap smear technique known as liquid-based cytology (LBC) was introduced in England and Wales (fig. 155). It preserved the cells in a liquid immediately after collection from the cervix and removed any blood or mucus before transferring the sample to a slide. The new technique, which produced clearer slides that were easier to examine, replaced the traditional Papanicolaou test and eventually became the standard method used throughout the UK. Previously one in ten tests had needed to be repeated because of problems with slide preparation. The development of LBC enhanced the efficiency and effectiveness of cervical screening, with the Department of Health pilot study showing that it facilitated an 87 per cent reduction in the number of tests that had to be repeated.[27]

CANCER OF THE CERVIX: A SEXUALLY TRANSMITTED INFECTION?

Since the introduction of the national cervical screening programme in 1964, cervical cancer mortality rates had been in decline. But to eradicate the disease, we needed to know the cause of cervical cancer. It was long suspected that cancer of the female reproductive organs was somehow connected with sexual intercourse. However, the idea was based only on anecdotal evidence and sociocultural assumptions. The link began to be investigated systematically using scientific methods as early as 1840, when the Italian surgeon Domenico Rigoni-Stern (1810–1855) used statistical evidence to illustrate that in Verona, 'cancer of the uterus' was more common in married women than in virgins and nuns. His data suggested that there may truly be an association between 'cancers of the womb' and sexual activity.[28] By targeting the married woman, Rigoni-Stern's findings complicated contemporary assumptions that cancers of the womb were associated with promiscuous sexual activity and moral laxity. This may explain why his findings were largely ignored during his lifetime.

There were a number of related studies over the following century, but little progress was made. During the 1960s, far-reaching social change, including the introduction of the oral contraceptive pill and the decriminalisation of homosexuality in 1967, encouraged a climate of greater sexual liberation, particularly amongst Britain's youth. Correlations between cervical cancer rates and sexual activity were exacerbated by this sexual revolution. Consequently, the 1970s saw a renewed interest in uncovering the aetiology of cervical cancer. In 1974 the epidemiologist Valerie Beral published a landmark paper in the *Lancet*, which analysed mortality rates from cervical cancer in successive generations of women born between 1902 and 1947 in England and Wales, and between 1902 and 1942 for Scotland. She found that mortality rates fluctuated in tandem with rates of gonorrhoea incidence, and therefore concluded that 'it has been established that sexual activity is a major factor in the genesis of cervical cancer'.[29]

Shortly afterwards, the German virologist Harald zur Hausen and his researchers began to investigate the root of this sexual activity factor. They began by searching for traces of herpes simplex type 2 (HSV-2) in cervical cancer biopsies. However, no significant correlation was found. Next, in the early 1980s, Hausen and his team moved their focus onto a sexually transmitted virus that causes genital warts – Human Papilloma Virus (HPV). They managed to isolate several different strains of HPV and found that certain strains were present in 93 per cent of the cervical cancer samples in their study. In 2008 Hausen was awarded the Nobel Prize in Physiology or Medicine for discovering Human Papilloma Viruses that cause cervical cancer. To date, this is one of just a few examples of a cancer being conclusively connected with a source, providing a definitive target for treatment.

Building on Hausen's research, a series of studies in the 1990s showed that virtually *all* cases of cervical cancer contained HPV of the same types.[30] By 2018, 100 different types of HPV had been identified, many of which were found to be completely harmless. At least 13 types of HPV were known to cause cancer, with most cases of cervical cancer being caused by two types, HPV 16 and HPV 18.[31]

AN EPITAPH FOR CERVICAL CANCER?

In 2008 a national HPV vaccination programme was introduced in Britain for girls aged 12 to 13 (fig. 156). This new vaccine was designed to prevent contraction of the most prominent strains of HPV that cause cervical cancer. Provided that a girl had not become sexually active by the time she received the vaccine, her chances of developing cervical cancer were drastically reduced by the HPV vaccination programme.[32]

The NHS justified the government's decision to vaccinate only girls, and not boys, against HPV, explaining that 'Vaccinating girls helps to indirectly protect boys from these types of HPV through what's known as herd immunity because vaccinated girls won't pass HPV on to them'.[33] This decision clearly located young female bodies as the source of infection, despite the fact that HPV is a gender-neutral condition. While men cannot develop cervical cancer, they can contract HPV and there are a wide range of HPV-linked cancers that do affect males, including cancer of the anus, penis, mouth and throat. Oral cancers are one of the fastest rising types of cancer worldwide and are more common amongst men. In the UK, rates of oral cancers have increased by 68 per cent in the past 20 years.[34] Today, in 2019, more than 80 per cent of the general adult population in Britain have HPV – making it the most prevalent sexually transmitted infection in Britain. By choosing to offer the HPV vaccination programme only to young girls, the government knowingly denied men the opportunity to reduce their risk of developing certain strains of cancer, particularly those who have sex with men, for whom 'herd immunity' originating with vaccinated women and girls is irrelevant.

The HPV vaccination programme was extended to men aged 45 or younger[35] who have sex with other men in Wales from 2015, in Scotland from 2016 and in England from 2018.[36] Transgender women can receive the vaccine at the discretion of a GP, based on 'a risk assessment that includes the woman's sexual behaviour and the sexual behaviour of her partners'.[37] In July 2018, it was announced that the HPV vaccination programme would be extended to boys aged between 12 and 13 in England. This was a triumph for healthcare equality, following a decade of campaigning

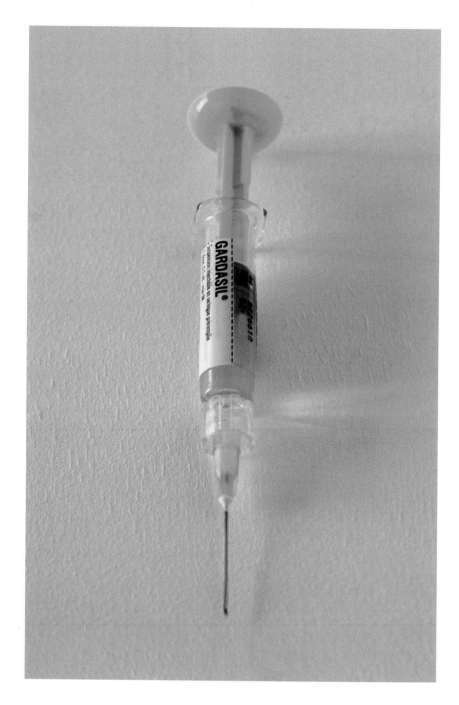

Fig. 156
The Gardasil vaccine helps protect against four of the most prominent strains of HPV that cause cervical cancer.

Fig. 157 (opposite)
The 'Easy' Girl-Friend, public health campaign poster for the Ministry of Health, designed by Reginald Mount, 1943–4.

Wellcome Collection

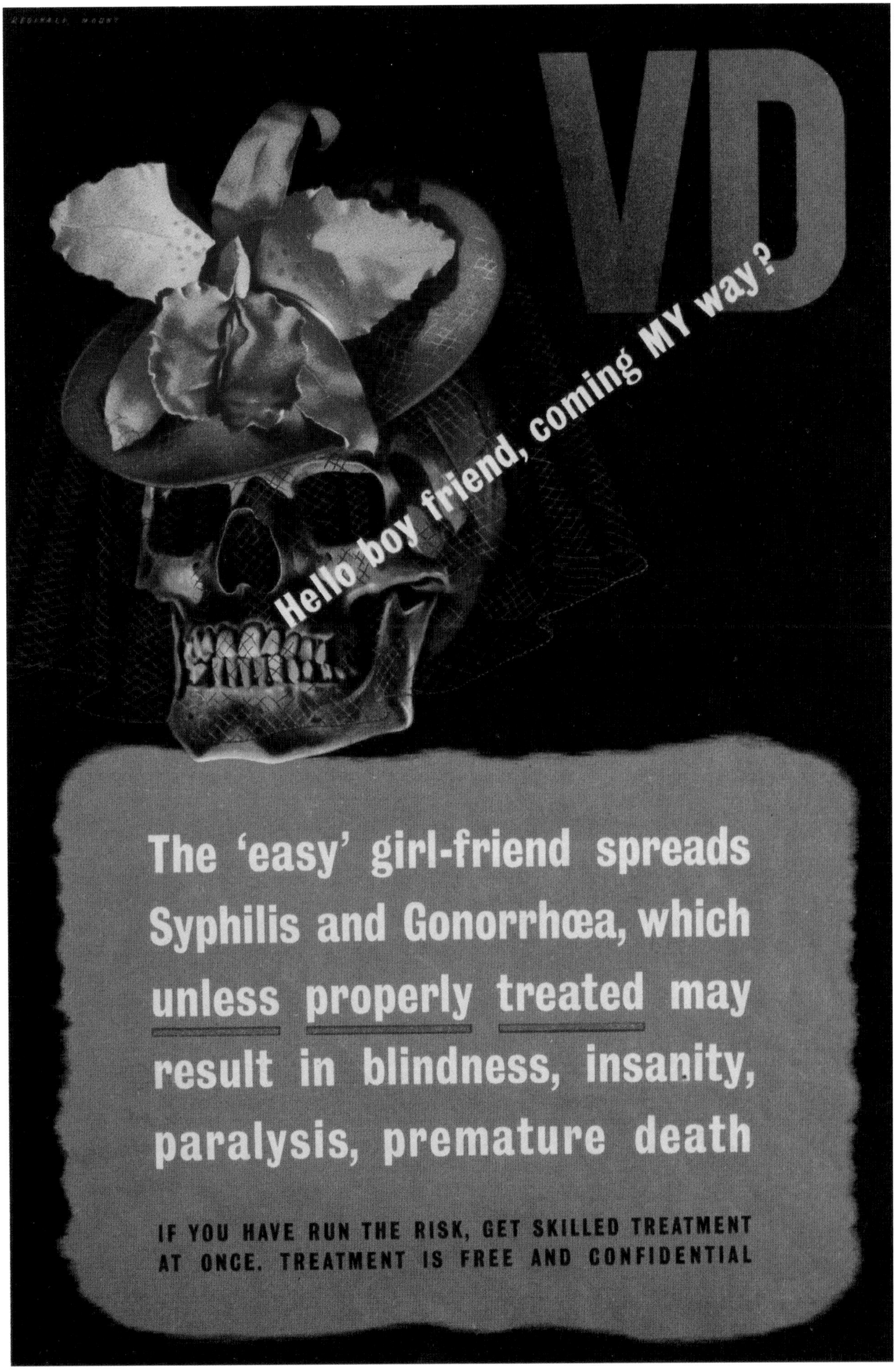
VD
Hello boy friend, coming MY way?
The 'easy' girl-friend spreads Syphilis and Gonorrhœa, which unless properly treated may result in blindness, insanity, paralysis, premature death
IF YOU HAVE RUN THE RISK, GET SKILLED TREATMENT AT ONCE. TREATMENT IS FREE AND CONFIDENTIAL

Fig. 158
A weight-loss
advertisement on public
transport that shows a
sexualised and objectified
image of a woman,
London, 2015.

and negative recommendations from the expert scientific advisory committee that guides the UK government on matters relating to vaccination and immunisation.[38]

The feminisation of HPV echoes the location of syphilis and gonorrhoea in the body of the 'promiscuous woman' in the 19th century (fig. 157). These examples are part of a long tradition of state and medical establishment transcribing social constructs onto female, and feminised, bodies. From the contraceptive pill and HIV/AIDS to abortion and in-vitro fertilisation, genitourinary and reproductive medical technologies are inseparable from the sexual politics of the society in which they are produced. The pathologisation of the female body is just one expression of the hyper-sexualisation of the female body in 21st-century capitalist society. From advertising and the media to pornography and social media, film and television to music and gaming, significant money and power are implicit in the construction of an idealised, sexualised female body (fig. 158). While the 'respectable' Victorian woman was controlled by a strong denial of her sexuality, positioned

in opposition to the diseased prostitute, the early 21st-century woman faces intense and pervasive sexualisation.

For those women who fell outside of the catchment age bracket when the HPV vaccination was introduced, regular screening remains the primary means of cervical cancer prevention. The HPV vaccination does not provide protection from all cancer-linked strains of HPV, and therefore all vaccinated girls must also attend cervical screening from the age of 25. In England, Wales and Northern Ireland, cervical screening now includes tests for changes to both cervical cells and HPV. This free, five-minute test should prevent most instances of cervical cancer from ever developing. And yet cervical cancer rates in women aged 25 to 49 have been rising since the early 2000s. Indeed, between 1993 and 2015 incidence of the disease increased by 95 per cent among women aged 20 to 24, and by 47 per cent among women aged 25 to 34. Cervical cancer remains the most common cancer among women under the age of 35.[39]

A 2017 study revealed a key cause of rising cervical cancer rates: one in four women in

England do not attend cervical screening when invited by their GP. This number rises to one in three women aged 25 to 29 and one in two in some deprived regions of the country.[40] A subsequent study surveyed 3,002 women in an attempt to understand the reasons why so many are not attending cervical screening. A third of participants said that embarrassment had prevented them from attending. This embarrassment was caused by their weight or body shape (35 per cent of the full sample and 50 per cent of participants aged 25 to 35), the appearance of their vulva (34 per cent) and concerns over smelling 'normally' (38 per cent). The study revealed that 31 per cent of participants aged 25 to 35 would not attend screening if they had not waxed or shaved their pubic hair. Other reasons cited included preferring not to know if there is something wrong (20 per cent), worrying that the test would be painful (25 per cent) and feeling uncomfortable removing their clothes in front of a stranger (25 per cent).[41] It would appear that the social construction of women's bodies is being internalised by women of all ages, creating a crisis of body shame so powerful that, at best, it is preventing women from caring for their health and, at worst, it is responsible for rising mortality rates from cervical cancer in Britain today.

Today, cervical cancer screening will once again metamorphosise. In England, the search for abnormal cells that was established by Papanicolaou in the 1920s will no longer be the primary test for the prevention of cervical cancer in the UK. Instead of searching women's cervical samples for signs of cellular changes manually through a microscope, cervical samples will undergo a DNA-based test for the presence of HPV. Only if a woman tests positive for a cancer-linked strain of HPV will her cells be examined for signs of cancer or pre-cancer. This is a major shift behind the scenes with machines (DNA sequencers) replacing many people (laboratory technicians). The patient experience of cervical screening will not change: a sample will be taken from the cervix in the same way as before. HPV primary testing represents a major technological and structural change for cervical cancer screening. However, evidence suggests that the dwindling UK cervical screening programme is due to low attendance at testing – not the efficacy of the test itself. Logic suggests that the change to the type of testing will therefore not affect the core issue that is driving cervical cancer mortality. From 2020, the oldest of the girls to first receive the HPV vaccination will reach the age of first HPV primary screening and we should begin to see the impact of the vaccination programme – if the young women attend.

Exploration of the intersection between gender and medical technologies can provide illuminating new insights into both phenomena, as the social scientists Maria Lohan and Wendy Faulkner suggest: 'since technology and gender are both socially constructed and socially pervasive, we can never fully understand one without also understanding the other.'[42] This is a useful approach for historians of medicine, technology and gender studies. But more importantly, this intersection has had a powerful impact on the lives, and deaths, of men and women for centuries – and this continues into the 21st century. The social forces that are preventing women from attending cervical cancer screening today are part of the same hegemonic framework that enabled prostitutes to be examined, imprisoned and treated by law in the 19th century, that allowed Mary Papanicolaou's contribution to the development of the Pap smear test to go unacknowledged in the mid-20th century and that targeted recruitment of cheap female labour to carry out the national cervical screening programme in the 1960s. Until advisory committees, policy makers and medical professionals begin to consider the wider structural problems that affect society, and the ways in which their decisions shape and are shaped by people, as we do in the humanities and social sciences, it is likely that investment and medical intervention will have only limited impact on rates of disease.

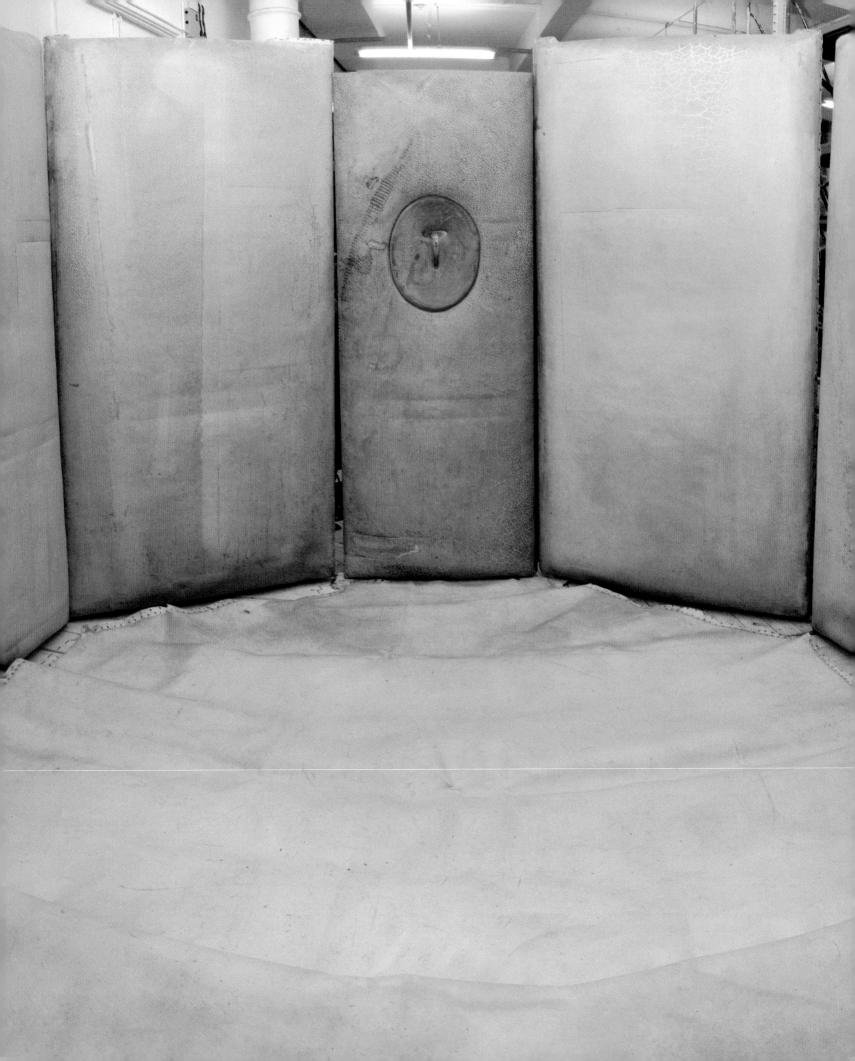

7
HEARING DISTANT VOICES
PSYCHIATRY, ASYLUMS AND THE LIMITS OF HISTORY

OISÍN WALL

Museum collections are like the fossil record. Only a tiny fraction of the world ever makes it in and, when it does, it is often at a moment of mass obsolescence. There are striking examples of this in the Science Museum's collections, which contain an unexpected array of non-medical-looking objects. A large wicker basket, farm tools, a grandfather clock, an ordinary kitchen chair, a wooden post box. Alongside other, more obviously medical, objects including electrotherapy machines, straitjackets and a full padded cell (fig. 159), they are the remnants of some of the great asylums that once haunted the outskirts of British cities – vast institutions often housing upwards of 2,000 patients, incorporating farms, workshops and laundries worked by the inmates. Babies were born in the hospitals and many patients died there and were buried in the grounds. Objects from the Science Museum's collection – a fossil record of the asylums – can help us to explore what happened to these institutions in the 20th century.

As with much of modern culture, the First World War is a good place to begin. Perhaps as a consequence of attempts to define and treat shellshock, a condition similar to modern post-traumatic stress disorder (PTSD), there was renewed interest in finding physical treatments in what were then still called 'asylums'. This was sparked by the need, on both sides of the War, to return soldiers to the battlefield and to identify 'malingerers', soldiers believed to be pretending to be ill to get out of fighting. However, it persisted through the next three decades, well

into the 1940s. At first, older therapies were resurrected. These physical treatments included therapies like Faradism, which involved running low-intensity electric currents through various parts of the body, despite the fact that this had already been tried during the late 19th century and abandoned as quackery. This initial exploration was followed by more successful treatments, like pyrotherapy, which was introduced in 1917 to treat soldiers with the final stages of syphilis, then known as 'general paralysis of the insane', by inducing high fevers; insulin coma therapy in 1927; chemical convulsive therapy in 1934; lobotomy in 1936; and electroconvulsive therapy (ECT) in 1938. The last of these became one of the most widely used psychiatric treatments of the mid-20th century. For many psychiatrists, ECT was the first treatment they would offer to most of their patients. Indeed, some thought of it less as a clinical decision and more as the patient's right. A report from the Cassel Hospital in West London in 1944, for instance, concluded: 'the question must not be whether [electroconvulsive] treatment should be administered but whether in a given case there is any legitimate reason for withholding it'.[1]

The particular electroconvulsive therapy machine pictured overleaf (fig. 160) was used by the Scottish psychiatrist Arthur Spencer Paterson (1900–1983). Paterson met Ugo Cerletti (1877–1963), one of the inventors of ECT, in Rome in 1946 and returned to the West London Hospital as a leading proponent of the technique. Indeed,

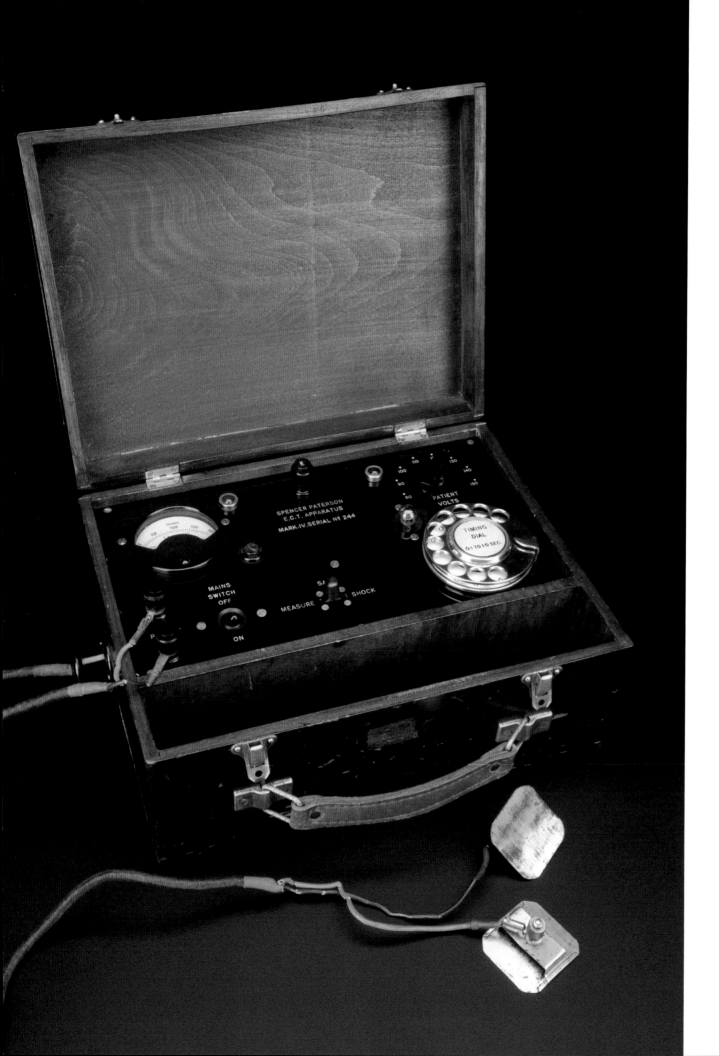

Fig. 160 (opposite)
Electroconvulsive therapy
(ECT) machine, designed
by Arthur Spencer Paterson
and manufactured by
Ferranti Ltd, UK, 1946–50.

Science Museum Group.
Object number 1984-159

Fig. 161 (right)
Patient receiving insulin
shock therapy.

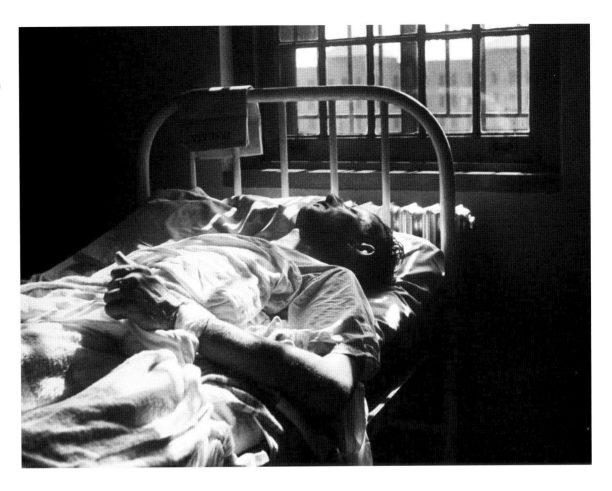

the machine in the Museum's collection is one of Paterson's own designs. Paterson saw ECT almost as a panacea and recommended it as a treatment for 'the anxious, the depressed, the deluded, the confused, and hallucinated'.[2] It worked by passing an electric current between two electrodes placed at the patient's temples, causing the patient to convulse. He described it as 'one of the most successful and least understood therapies in medicine'.[3] He was writing 20 years after it had first been used and 15 years since it had been widely accepted by psychiatrists – various research projects had explored the biochemical changes in the brain that followed a round of treatment and others speculated that it might work as a subconscious stress, or even as a punishment, but still there was no consensus about how the treatment actually worked.

Another common treatment was insulin shock therapy, which was usually used to treat people with schizophrenia (fig. 161). Developed in 1927, this treatment used insulin, like that used by people with diabetes to control their blood sugar, to induce a coma in the patient. Usually these comas lasted under an hour but, since there was

no widely accepted protocol for this treatment, the length of the coma depended on the doctor's judgement. Some psychiatrists also combined insulin comas with shock therapies, using ECT or drugs to induce seizures during the coma. Courses of treatment could last for months with patients being put into a coma daily. We can see examples of this from the records book at Brighton County Borough Mental Hospital (figs 162 and 163). On the second page there is a column marked 'No. of comas', which shows that one patient had received more than 20 coma treatments. The treatment was dangerous and needed well-trained staff and a carefully controlled environment, but in spite of the risks, many medical professionals believed in its worth, with some, like the well-known British psychiatrist William Sargant (1907–1988) claiming that it should be the first-line treatment for schizophrenia.[4]

The apparent success of treatments like these inspired intense therapeutic optimism from the 1930s until well into the 1950s. Many psychiatrists looked back on their 19th-century predecessors with pity or scorn at what they

PALMER. R.M. (Cntd on page 80)

DATE	UNITS	REACTION	TIME OF INTERRUPTION	TIME AWAKE	METHOD OF INTERRUPTION	NO OF COMAS	NO OF FITS	REMARKS
23.2.48	20	Nil	10.30		10 gms Glucose Orally & Breakfast			
24.2.48	35	Nil	10.40		20 Gms Glucose Orally + Breakfast			
25.2.48	55	Nil	10.40		40 gms Glucose Orally & Breakfast			
26.2.48	75	Nil	10.40		50 gms Glucose Orally & Breakfast			
27.2.48	95	Perspired freely	10.55		60 Gms Glucose orally + Breakfast			
28.2.48	110	Nil	10.40		20 Gms Glucose orally + Breakfast / 100 cc 33% Sol glucose. I.V.			
1.3.48	115	Sopor 11.0	11.15	11.19	60 gms orally + Breakfast / 100 c.c. 33% Sol glucose. I.V	1		Extremely Abusive & noisy early morne.
2.3.48	125	Sopor 10.42	11.38	11.41	60 gms orally Breakfast / 100 cc 33% glucose. I.V	1 - 2		Paraldehyde ʒii @ 6.45 a.m
3.3.48	135	Sopor 10.35 Coma 11.34	11.39	11.42	70 gms Orally & Breakfast	1 - 3		Paraldehyde ʒii @ 6.45 A.M.
4.3.48	140	Sopor Coma 10	10.25	10.29	gms orally + Breakfast / 100 cc Sol glucose I.V			Paraldehyde ʒii @ 6 a.m
5.3.48	140	Sopor 9 Coma 10.15	10.30	10.34	70 GM glucose orally. Breakfast.	1 - 4		Paraldehyde ʒii @ 6.45 am
6.3.48	70	Perspired freely	10.30	10.34	60 gms glucose orally Breakfast / 100 cc Sol glucose I.V	1 - 5		Paraldehyde ʒii @ 6.45 am
8.3.48	140	Sopor 10 Coma 10.40	11.0	11.3	40 gms orally + Breakfast / 100 cc 33% Sol glucose I.V	1 - 6		Paraldehyde ʒii @ 6.45 am
9.3.48	140	Sopor 9 Twitch 9 Coma	10.5	10.8	70 gms orally + Breakfast / 100 cc 33% Sol glucose I.V	1 - 7		Restless after Interruption Paraldehyde ʒii @ 6.45 am
10.3.48	135	Sopor 9 Coma 10	11.00	11.04	70 gms orally + Breakfast	1 - 8		Paraldehyde ʒii @ 6.45 a
11.3.48	130	Sopor 9.20 Coma 10.30	10.52	10.56	100 cc 33% glucose I.V. 70 gms orally.	1 - 9		Paraldehyde ʒii @ 6.45 am
12.3.48	135	Sopor 10.10 Twitch 9 Coma 10	10.52	10.55	180 cc 33% glucose I.V 70 gms orally I.V.	1 - 10		Paraldehyde ʒii @ 6.45 am
13.3.48	70	Sopor 10 Twitch 9 Coma 10	11.10	11.15	30 gms orally + Breakfast	1 - 11		Paraldehyde ʒii @ 6.45 am
15.3.48	135	Sopor 9.45 Coma 11	11.30	11.33	100 cc 33% glucose I.V. 70 gms orally.	1 - 12		Paraldehyde ʒii @ 6.45 am
16.3.48	135	Sopor 10 Twitch Coma 10.30	11.2	11.8	orally + Breakfast / 100 cc 33% glucose I.V.	1 - 13		Paraldehyde ʒii @ 6.45 am
17.3.48	135	Sopor 10 Twitch Coma 11	11.7	11.10	70 gms orally / 100 cc 33% glucose. I.V	1 - 14		Paraldehyde ʒii @ 6.45 am
18.3.48	135	Sopor 9.10 Coma 11	11.35	11.38	70 gms orally / 70 cc 33% glucose I.V.			
19.3.48	135	Sopor 9.22 Coma 10	10.50	10.53	80 gms orally + Breakfast / 100 cc 33% Sol glucose I.V.	1 - 15		
20.3.48	70	Sopor 10.45 Twitch 10.29	10.58	11.2	30 gms orally + Breakfast	1 - 16		Restless during interruption
22.3.48	135	Sopor 10.15 Twitch 10 Coma 10.40	11.5	11.8	100 cc 33% Sol glucose I.V. 70 gms orally	1 - 17		Paraldehyde ʒii @ 6.45 am
23.3.48	135	Sopor 10 Twitch 9 Coma 10	10.30	10.55	70 gms orally / 100 cc 33% glucose I.V.	1 - 18		
24.3.48	135	Sopor 9.35 Coma 10.10	10.20	10.23	70 gms orally + Breakfast / 100 cc 33% Sol glucose I.V.			
25.3.48	135	Sopor 10 Twitch 9 Coma 10	10.35	10.38	40 gms orally + Breakfast / 100 cc 33% Sol glucose I.V	1 - 19		
26.3.48	135	Sopor 9 Twitch 8.55	11.10	11.13	70 gms orally + Breakfast / 100 cc 33% Sol glucose I.V.			
27.3.48	70	Sopor 10.34 Twitch 10.10	10.55	10.58	30 gms orally / 100 cc 33% Sol glucose I.V			
29.3.48	135	Sopor 10 Twitch 9 Coma	11.15	11.30	70 gms orally + Breakfast / 100 cc 33% Sol glucose I.V.	1 - 20		
30.3.48	135	Sopor 9.50 Coma 10.45	11.5	11.8	70 gms orally + Breakfast			

Figs 162 and 163 (opposite)
Records book, Brighton
County Borough Mental
Hospital, 1946–8. The book
documented the use of
insulin and its effects on
male patients.

perceived as barbaric treatments or even outright neglect. Gone, they believed, was the time when asylum 'keepers', as they were called, were the gaolers of the mad. From 1930 onwards, the number of outpatients grew and in the 1950s the numbers of inpatients began to decline for the first time in a century. It seemed to many practitioners that real, practical treatments could be offered – cures, even – from which some patients would walk away never to need psychiatric care again.

However, these treatments were not without their detractors. There had always been sceptics, and from the 1950s onwards these voices began to grow louder. They argued that the treatments were violent, dehumanising, cruel and pointless. The critics offered new explanations of the success of these treatments. Some suggested that patients recovered because of the increased care, personal contact and attention they received before and after the treatment. Others argued that their efficacy was an illusion, created by patients too afraid of the treatment to admit any further mental health difficulties. Still others claimed that they simply did not work at all and that their perceived success was due to the deliberate selection of patients who already had a good prognosis.[5] By the late 1960s, these treatments were in decline across Britain, and a decade later, almost no hospitals practised insulin coma therapy, chemical shock therapy or lobotomy, and the use of ECT was far less common.

The range of new physical treatments developed in the years between the First and Second World Wars was not the only thing that changed psychiatry. In 1930 the Mental Treatment Act made the first provisions for the treatment of psychiatric outpatients; the 1940s saw the development of new group therapeutic approaches; the 1950s saw the first therapeutic hostels and halfway houses in Britain and the first decline in psychiatric inpatients since the beginning of the modern asylum system, as well as the invention of new antipsychotic drugs.

The first antipsychotic drug, Largactil, was introduced in 1953 (fig. 164). It was widely used as a treatment for psychotic disorders such as schizophrenia. However, while regular treatments were commonly given as a tablet or suppository, this drug was designed to be given with a hypodermic syringe. This meant that it could be delivered to patients who were seen to be 'acting out' – in distress or behaving in a disruptive way. Injecting the drug allowed staff to administer it

rapidly and without the cooperation, or indeed consent, of the patient. It is this practice that earned Largactil the moniker 'the chemical cosh' – a chemical alternative to the batons that previous generations of asylum staff carried to subdue unruly patients. As the first effective and reasonably sustainable treatment for schizophrenia, Largactil transformed the psychiatric landscape. It meant that many people with severe problems who might otherwise have spent their lives in a 'back ward', where long-term patients were housed, could be returned to their communities and live productive lives.

This was a source of new optimism, which, unlike that of the interwar years, was focused on the long-term management of symptoms, rather than a cure. By 1957, this optimism was already so entrenched and widely accepted that T. P. Rees, the President of the Royal Medico-Psychological Association (the forerunner to the Royal College of Psychiatrists) was able to say in his presidential address:

The other day an Infectious Disease Hospital closed down and was put up for sale, not for

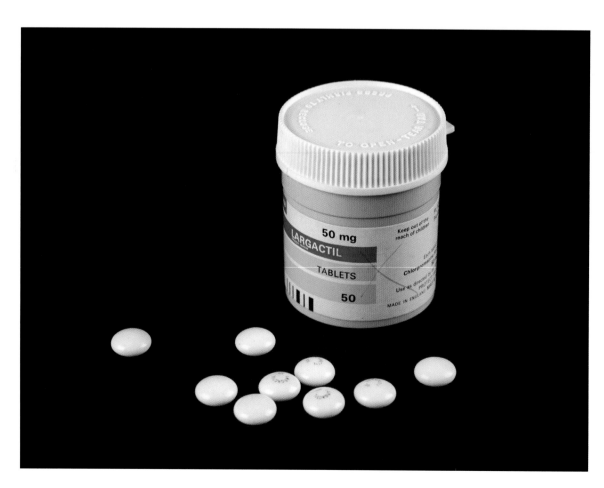

Fig. 164
Largactil in tablet form, the first widely used antipsychotic drug.

Science Museum Group.
Object number 1986-1248/530

lack of staff, but for lack of patients. Perhaps those who are coming after us will regard this period through which we are now passing in psychiatry as the mental hospital era, and maybe at the Annual Meeting of the Royal Medico-Psychological Association in 2056, a coach trip will be arranged to pay a visit to the last of the big custodial care long-term mental hospitals before it is finally converted into a holiday camp![6]

Rees's comments were prescient. In the three decades following his speech, the number of psychiatric inpatients fell 97 per cent, from 300,000 to 10,000. In the 1960s, psychiatric optimism coincided with a series of governments committed to cutting the cost of psychiatric services. By the next decade, this drive was added to by an increasingly organised campaign by groups such as the Mental Patients' Union, which campaigned for psychiatric reform.[7] During these years, the large hospitals were systematically taken apart. The farms and workshops were sold off, inpatient numbers reduced, wards closed, and new outpatient facilities and hostels were opened in surrounding communities. In 1983, four years after she came to power, Margaret Thatcher introduced a policy known as 'care in the community,' which accelerated the process that became known as 'deinstitutionalisation'.[8] Between 1983 and 1996, most of the psychiatric hospitals in Britain were closed. Patients were moved into hostels, therapeutic communities, general hospitals and even into the private homes of families paid to take them in. However, this acceleration's lack of proper support or funding meant there was not always appropriate accommodation available for discharged patients, and social workers were overworked trying to care for people who sometimes had little life experience outside the regimented world of the hospital. The communities that people with mental health difficulties were moved into were often unprepared and unwelcoming. As a result of these shortfalls, some people were marked for the rest of their lives by stigma.

All of these objects – the ECT machine, the insulin therapy book and the drugs – tell a particular kind of history: the history of a profession, of institutions, of middle-class men experimenting on, and exercising power over, other people. However, this is only one side of the history of psychiatry. There have always been more patients than staff within the psychiatric system and, although their voices are just as important, they are often harder to hear. When one wants to find out about the experiences of psychiatrists, such as T. P. Rees or A. S. Paterson, one can look to countless published papers by and about these men, but recovering the experiences of their patients can be almost impossible. As a result, the lives of 'great men' can sometimes dominate the way we tell the history of psychiatry. One of the great strengths of the modern discipline of history is that it embraces techniques that can disrupt these received narratives. The study of 'material culture' is one such technique.

While the objects we have discussed feed into received narratives, one of the important features of a collection like that of the Science Museum is that it also contains everyday objects not associated with any grand historical actors. They provide us with fragments, enough to remind us that there was once a story, but not enough to tell us what that story was. In many ways, such objects lead us to a more personal engagement, calling out to us to imagine their purpose, their relationship to donor or patient, to think about the lives gladdened or saddened by them, and the fragility of our relationship to the world of things. These objects remind us of the patients, porters, visitors and others whose voices are recorded so rarely, and remind us that their stories have, in many ways, been elided from the historical record. The objects allow us to remember them in the unanswered and unanswerable questions they raise.

A birdcage (fig. 165) came to the Museum with a label recording its donation, along with 'four fine parrots,' by one Major Meek to the women of the Brighton County Borough Asylum. But who was Major Meek and why did he choose to give caged birds to asylum inmates? Was he a friend, brother, son, father or husband of one of them? Fig. 166 is a key, fashioned from the back of a piece of cutlery and confiscated by a member of staff at St Francis' Hospital in Brighton. But what lock was the key made to open, and did it work? Did its maker use it to travel freely and secretly within the confines of the hospital, or even outside? Or did it merely betoken freedom to the person who so carefully fashioned it? Another item in the collection (fig. 167) is a painting by a patient at St Audry's

Hospital in Suffolk (fig. 168), who recorded their life there. Looking at it, one wonders who the people behind the first-floor windows of the hospital are. Whose room lay behind the densely barred, partially open window on the second floor?

These are objects that pose unanswerable questions, that remind us of long-forgotten narratives of the grievous pain, banality and tragedy of the patients' lives that underpin the history of psychiatry. They do not fit neatly into the apparently straightforward march of history that led from custodial asylums to care in the community. They represent the lives and stories excluded and unspoken in this history. These objects were collected by the Science Museum from closed hospitals, sometimes purchased, donated, or, as a last resort, pulled from skips. Where possible the curators collected contextualizing documents and testimonies, but some important objects came with little or no context or explanation, isolated stones to trouble the flowing narrative of science. It is because of this absence that they represent the voices of people who have been excluded from history, and the limits of what a historian can do.

Fig. 165 (below left) Birdcage donated, with 'four fine parrots,' to the female patients of the Brighton County Borough Asylum.
Science Museum Group. Object number 1996-271/31

Fig. 166 (below right) Key made from the back of a piece of cutlery, by a patient at Brighton County Borough Mental Hospital/ St Francis Psychiatric Hospital, 1900–60.
Science Museum Group. Object number 1996-271/4

Fig. 167 (opposite above) Painting of St Audry's Hospital, Suffolk, made by a patient during their time at the hospital.
Science Museum Group. Object number 1990-183/18

Fig. 168 (opposite below) Aerial photograph of St Audry's Hospital, Suffolk.

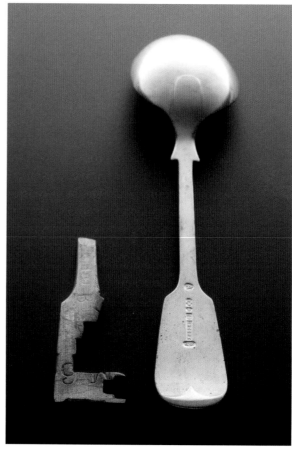

8
THE PHARMACY SHOP
'A GREAT RAGE FOR MAHOGANY, VARNISH AND EXPENSIVE FLOOR-CLOTH'

BRIONY HUDSON

Fig. 169
The apothecary John Simmonds and his apprentice William, working in the laboratory behind John Bell's pharmacy, 1842.

Wellcome Collection.
Object number 1997-375

With its 'empty boxes, green earthen pots, bladders and musty seeds', Shakespeare's account of a 17th-century apothecary in *Romeo and Juliet* conjures up an evocative image of a familiar, yet exotic and ancient, trade. Today, the pharmacist's role is so bound up with retail business that many people would be hard pressed to remember that they also work in significant numbers in hospitals, not to mention in industry and academia. For community pharmacists and their apothecary predecessors, the shop is, and was, a defining characteristic. When the founder of the Pharmaceutical Society of Great Britain (PSGB), Jacob Bell (1810–1859), wrote his father's biography, his focus was obvious: 'towards the end of 1798, John Bell opened his shop'.[1] The shop at 338 Oxford Street, London, was extremely successful, surviving today as John Bell & Croyden, Lloyds Pharmacy's flagship store on Wigmore Street. John Bell also played a part in establishing the PSGB in April 1841. But for Jacob, recounting his father's career, the most obvious milestone was the opening of an independent premises to sell and prepare pharmaceutical products: a shop (fig. 169).

This focus on the shop has always been a mixed blessing for pharmacy. The taint of trade causes reputational issues, even today. The struggle to establish a pharmacy profession by the Victorian chemists and druggists surrounding Jacob Bell was exacerbated because they were not taken seriously as educated science-based practitioners. The aura of quackery was hard to shake off when accusations of prioritising profit over

altruistic healing were so easily presented by the opposition (usually university-educated physicians, or examined apothecaries with the weight of their livery company behind them). Today's community pharmacists contend with the reality that many customers are buying their lunchtime sandwich or picking up toothpaste, rather than benefiting from their significant expertise in pharmacology, dosage delivery systems and adverse drug reactions. Some customers' impatience to wait for their prescription illustrates their misunderstanding of the pharmacists' clinical role.

But the shop also has its benefits. Promoted by the Royal Phamaceutical Society in recent years as 'the scientist on the high street',[2] the pharmacist has always been visible and accessible in a way that a doctor or surgeon is not (fig. 170). And for historians and museum curators, the sheer quantity of items that a pharmacy uses, displays and produces means they leave a significant trace in collections. While other medical professions are often represented in exhibitions by a limited range of instruments or portraits, the surviving packaging, containers, advertising and equipment means that the retail pharmacy's impact over time can be experienced. All of these are items that were designed and devised to elicit an emotional response, and even today, a 19th-century advert for teething powder or a row of delftware drug jars can evoke the same reaction in a modern viewer as in the Victorian customer. It gives us a real sense of how our historical counterparts were persuaded

Fig. 170
Heppel & Co. Chemists,
Strand, London, 1912.
This 24-hour pharmacy
housed an overwhelming
quantity of stock, with
brushes, sponges and
perfumes, and even an
American soda fountain
amongst the medicines.

Fig. 171
Gibson & Son's
Chemists and Opticians,
reconstruction in the
Science Museum's former
Medicine Galleries,
1980–2015.

to put their trust in the pharmacist and part with their money.

An example of this is the reconstruction of Gibson & Son's shop, run by the Gibson family in Hexham, Northumberland, from 1834 until its closure in 1978, that now forms the centrepiece in the Science Museum's Medicine Galleries' presentation of pharmacy history (fig. 171). This chapter explores the British pharmacy shop as a reflection of the relationship between the seller of medicines and their customers, the perception of the apothecary/pharmacist and the evolving range of products, medicinal and otherwise, available to the public over time.

'GREEN EARTHEN POTS, BLADDERS AND MUSTY SEEDS': THE APOTHECARY'S SHOP, 1610s TO 1760s

The predecessor to today's community pharmacist was the apothecary, or at least this was so in the 17th century. Having previously formed a sub-group in the Grocers' Livery company, the Worshipful Society of Apothecaries was granted its royal charter in 1617, establishing a specialised body to represent, regulate and develop a membership of practitioners with pharmacy skills. Like other livery companies in the City of London, its role included oversight of apprentices progressing through the trade and, in its specific pharmaceutical guise, it gained power to search apothecaries' shops to inspect the quality of their drugs, to ensure the highest standards and maintain the trade's reputation. When William Shakespeare described an apothecary's premises in *Romeo and Juliet* (1597), its owner is 'in tattered weeds' in a 'needy Shop' stocked with 'a beggarly Account of ... remnants of Packthread and old Cakes of Roses' (fig. 172). Here was a retailer struggling to make a living, his poor circumstances linked to his willingness to supply poison to the doomed lovers.

Unlike Shakespeare's fictional Italian apothecary, many British apothecary shops in the 17th and 18th centuries were vibrant places, acting literally as a window on the medicines available to those who could afford them, and increasingly embracing exotic global discoveries alongside more traditional remedies and recipes. In London, apothecaries clustered around Cheapside and Bucklersbury from the 14th century onwards.

A Venetian visitor to the city, around 1618, noted that this area was 'full of apothecaries' shops on either side of the way'.[3] The Society of Apothecaries' reach was only seven miles from the City, but there were also many people carrying out similar roles across the country. For example, the historian Alan Withey has found at least 124 identifiable apothecaries in Welsh records in the period 1600 to 1762.[4]

Before 1650, the rise of new apothecary shops as highly specialised retailers reflected the sheer number of new imported drugs available, including aloes from China and Borneo; mercury from Tibet; camphor from Sumatra and Java; myrrh from the Arabian and African coasts; and senna, rhubarb and sarsaparilla from the Americas. An inventory of Robert Baskerville's Exeter shop in 1596 shows very early listings of nux vomica (native to India) and sassafras (found in North America and East Asia).[5] In June 1667 Samuel Pepys's wife, Elizabeth, was 'making of Tea, a drink which Mr Pelling the apothecary tells her is good for her cold and defluxions'.[6] Although cinchona was established as a remedy in England by the 1670s, John Quincy wrote in 1718, 'This Simple is so lately brought into Medicine, that there is little to be met with in Authors about it; and People's Notions seem yet so confus'd and undetermin'd concerning its Virtues and Efficacy.'[7] Exotic goods were flowing into the country so rapidly that medical usage was struggling to keep up.

For centuries, people had turned first to their own remedies to treat illness, often made up from hedgerow ingredients. This practice did not change overnight as imported substances became available. Surviving records show that 'kitchen-physic' continued and was encouraged by published recipe books. A spectrum of remedies existed from established native plants, through recipes that combined local and shop-bought ingredients, to a wide range of ready-made products created by the apothecary including cordials, vomits, purges and pills. In 1684 the Swansea apothecary Richard Philpots's stock included 'Spanish white' [wine?], capers and anchovies, sugar and common turpentine, 'syrops, electuarys and oyles', and 'chimicall preparations and salts'.[8] Clearly these were a mixture of medicaments made up on site, available alongside ingredients (not all purely medicinal) that could be taken home.

THE
APOTHECARY

W. Shakespear Inv.

Dr Rock Sculp.

I do remember an Apothecary,	An Alligator ſtuft, and other Skins
And hereabouts he dwells, which late I noted	Of ill ſhap'd Fiſhes, and about his Shelves
In tatter'd Weeds, with overwhelming Brows,	A beggarly Account of empty Boxes;
Culling of Simples; meager were his Looks,	Green earthen Pots, Bladders, and muſty Seeds,
Sharp Miſery had worn him to the Bones:	Remnants of Packthread, and old Cakes of Roſes
And in his needy Shop a Tortoiſe hung,	Where thinly ſcattered to make up a ſhew.

Accord.^g to y.^e Act. To be had of T. Ewart facing Old Slaughters Coffee Houſe St Martins Lane Long Acre. Price 6.^d

BOVTIQVE PHARMACEVTIQVE.

The Apothecary's Shop opened.

Sold by N: Brooke at ye Angell in Cornhill.

Fig. 172 (opposite)
Dr Rock, *The Apothecary*, engraving, c. 1750. Inspired by Shakespeare's description of the apothecary hard at work, surrounded by equipment, exotic goods hanging from the ceiling and a jar of leeches in the window.

Fig. 173 (top)
Jean de Renou, *Les oeuvres pharmaceutiques* (detail), c. 1608. Renou, a pharmaceutical authority and chief physician to the French king, wrote this authoritative work on correct practices.

Wellcome Collection.

Fig. 174 (above)
Michel Morel, *The Expert Doctor's Dispensatory*, 1657. The frontispiece shows the apothecary settled behind the counter, tending to a customer.

Wellcome Collection

Sources are scarce that detail apothecary shop exteriors in this early period (fig. 173). When Cornelius Lyde, a druggist, became the lessee of the 'Black Lyon' site on Fleet Street in June 1729, it had 'two large sash windows in front glazed with crown glass run double, outside shutters and iron pins and keys, a sash door with a shutter with a large H hinges and spring lock and two iron bolts and brass nob latch, a Gothick sash over the door, a Tuscan frontice piece with a pediment and intablature covered with lead.'[9] Clearly security and aesthetics were both key concerns.

The most detailed insight into the interior of London apothecaries' shops is through inventories, including some now belonging to the Guildhall Library and London Metropolitan Archives, which have been investigated by the historian Patrick Wallis.[10] The quantity and variety of drugs is startling: two apothecaries each held more than 200 different substances. Apothecaries also stocked spirits, dyes, artists' pigments, chocolate, perfumes and tobacco, only some of which had any medicinal value. These blurred boundaries meant that apothecaries were not solely medical traders. The extremely detailed inventory of Thomas Needham's shop in Chesterfield in 1666 shows that he stocked 'marmalade, dryed Apricocks, dryed peares, candied ginger, machpane, comphits, makaroons' alongside chemicals including white lead, antimony, and arsenic.[11]

By the mid-17th century, apothecaries' shops had developed a distinct style with rows of ceramic jars on extensive runs of shelving (figs 174 and 175). There was a practical need for significant storage for a dizzying number of jars, pots, boxes and barrels. In 1669 Arthur Coldwell, an apothecary in Oundle in Northamptonshire, shoehorned storage into every available space. Customers entered the shop to see a wall covered with shelves bearing 70 gallypots, 35 small boxes, 12 barrels, a box of rice and a parcel of little glasses and chemical oils. Old drugs and spare bottles and pots were stored on a shelf above the door.[12] Wallis's analysis reveals that a significant proportion of apothecary shops' value was represented by fixtures and equipment, and that expenditure on decorative jars was without parallel in other retail trades.[13]

The rows of drug jars fulfilled a purpose, but there was no practical reason for their high level of decoration: plain ceramics and cheaper

wooden boxes or drawers could and did store pharmaceutical goods. Wallis argues that the drug jars established a separation between the raw material and the finished product, in order to highlight the apothecaries' labour and the authenticity and reliability of the drugs. They were therefore part of the apothecaries' promotional technique (fig. 176). Not every consumer felt comforted by their value. In 1670 the physician Jonathan Goddard observed, 'it is little wonder so many young apothecaries set up ... anew, and open shops in every corner almost of the City [of London], for it requires no great sum to purchase fine painted and guilded Pots, Boxes, & Glasses; and a little stock is improveable to a manifold proportion of what is capable of in other trades'.[14] Could inexperienced or duplicitous apothecaries prioritise style over substance?

Glass containers were less common than ceramic, partly due to their expense. Nevertheless, Thomas Baskerville's shop in Exeter in the 1590s contained gallon, pottle [two quarts] and quart 'glasses' or bottles, which were in this period used primarily for storing distilled waters. In 1637 the London shop of the apothecary John Arnold contained 117 'glasses' of different kinds, as well as 295 pots and jars, and 183 boxes and barrels.

Another distinctive symbol of the apothecary shop was exotic animals and fruit hanging from the ceiling. In Shakespeare's apothecary's premises 'a tortoise hung, An alligator stuffed, and other skins Of ill-shaped fishes' (see fig. 172).[15] The London apothecary Thomas Prescott listed a stuffed alligator in his 1686 inventory, and Thomas Johnson showed the first bananas to reach England in his shop window in the spring of 1633. Charles Duck displayed a 15-foot-long serpent in his shop in Kent. These displays of natural history suggested luxury, mystery and also scholarship linked with the growing public interest in natural philosophy. An apothecary's reputation could be coupled to the idea of informed scientific enquiry and debate (fig. 177).

In spite, or perhaps because, of the level of success achieved by many apothecaries in this period, their shops were designed to reassure as well as attract customers, since there was some mistrust. Apothecaries were sometimes subjected to public attacks over poisoning, excessive profiteering from exotic ingredients and exploiting

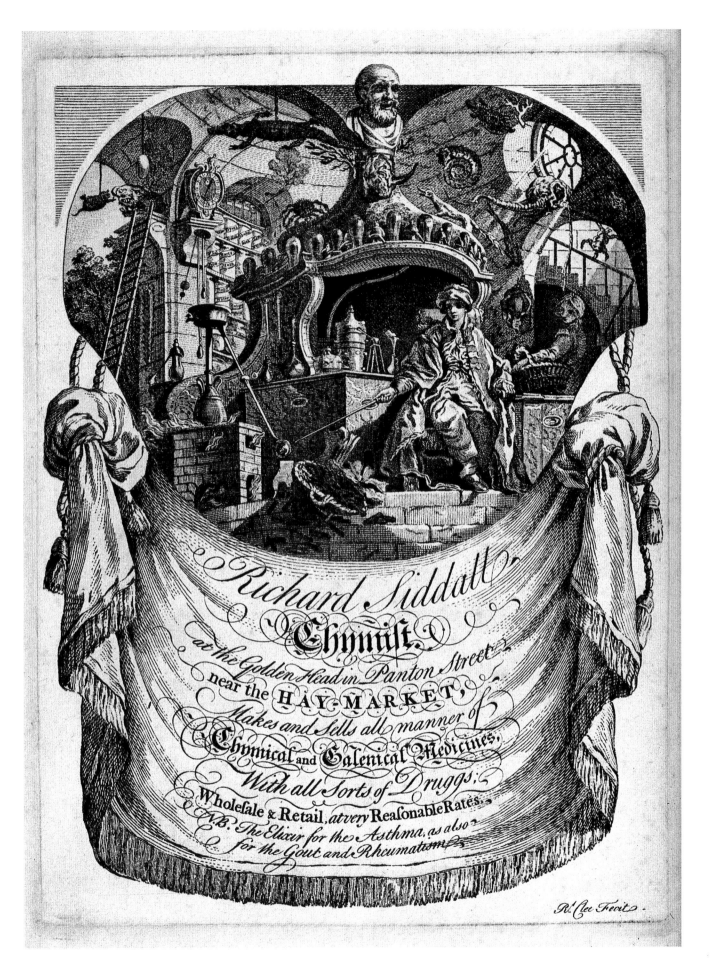

Richard Siddall

Chymist

at the Golden Head in Panton Street,
near the HAY-MARKET,
Makes and Sells all manner of
Chymical and Galenical Medicines,
With all Sorts of Druggs;
Wholesale & Retail, at very Reasonable Rates.
N.B. The Elixir for the Asthma, as also
for the Gout and Rheumatism.

R. Clee Fecit.

the sick and dying. Christopher Merrett, writing as a successful physician in 1670, questioned the apothecary's motives by commenting that they should spare the patient the 'charges of leaf-gold for gilding pots, glasses, pills, electuaries, boles &c, which serve only to raise the bill.'[16]

'I HAVE A SECRET ART TO CURE': MEDICINES AND SHOWMANSHIP, 1770s TO 1830s

Changing technology and retail methods combined with a massive acceleration in consumer spending were to have a significant impact on pharmacy shops from the late 1700s. This era saw the rise of the chemist and druggist, a specialist in the preparation and sale of medicines, in contrast to the apothecary who, by now, operated primarily as a general practitioner with a focus on diagnosis rather than medicines. The shop was a chemist and druggist's defining characteristic, and one that was growing in number. For example, Sheffield's first druggist business was recorded in 1750. There were ten

in 1797, 38 by 1838 and 56 by 1841. This was a national trend: three chemists in Bristol in 1775 grew to 114 in 1851; 13 in Nottingham in 1806 grew to 63 in 1850; two in Merthyr Tydfil in 1822 grew to 16 in 1850. Apothecaries abounded in the countryside as well as in urban areas (fig. 178).

In parallel, there was growth through the 18th century in the manufacture and distribution of commercially promoted proprietary medicines. Names such as 'Daffy's Elixir', 'Dr James's Fever Powder' and 'Dalby's Carminative' were each originally formulated by a single man, but grew quickly to be household names, marketed vigorously (fig. 179). Pre-packaged products were a novelty compared to home-prepared or made-to-order medicines, but the consumers were there, owing to the increasing number of waged jobs. In addition, developing transport networks, and especially the rise of the railway system, meant that supply could meet demand.

Patent medicines were a new development, but traditional products still survived. Adverts for John Toovey on the Strand in London show that he stocked products prepared on the premises

Fig. 178
M. Darly, *Matthew Manna, a Country Apothecary*, etching, 1773. The sign outside Manna's shop advertises his services as 'Man midwife. Gentlemen shaved. Hogs gelded. Shave for a penny & bleed for 2 pence'.
Wellcome Collection.

Fig. 179
The apothecary in this advertisement promotes his 'Paris Pill' and 'Balsamick Electuary', late 17th century.
Wellcome Collection.

MATTHEW MANNA. A COUNTRY APOTHECARY.

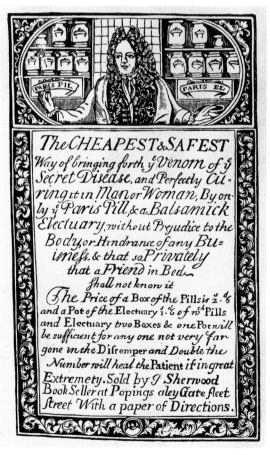

alongside others bought in, both wholesale and retail. He prepared 'all sorts of Chemical and Galenical Medicines ... the very best French and English Hungary Waters, Lavender and Mineral Waters, Daffy's and Stoughton's Elixir, etc. Wholesale and Retail ... Physicians Preparations made ... Chests of Medicines for Gentlemen and Exportation'.[17] Like its forerunners, this was still a retail business covering a broad base (fig. 180).

Shop signs extending over the street were prohibited in London after the 1760s. Not only did they crowd the street and occasionally fall, with disastrous results, but the introduction of numbering systems and increased literacy made them outdated. Some retailers expanded with bow windows: the apothecary Matthew Hinton was reprimanded by the Chester city authorities in 1767 for projecting his shop window into Lower Bridge Street. By the 1830s, Charles Dickens noted that linen-drapers and haberdashers were the first to embrace new technology with an 'inordinate love of plate-glass, and a passion for gas-lights and gilding ... [it then] burst out again amongst the chemists; the symptoms were the same, with the addition

of a strong desire to stick the Royal Arms over the shop-door, and a great rage for mahogany, varnish and expensive floor-cloth.'[18] From the 18th century, mahogany 'drug run' drawers displayed an impressive breadth of available products, with their gilded labels providing a decorative stock-list (fig. 181).

Window displays were an opportunity to make a shop distinctive, and, for chemists, the items to be included were carboys, large globular glass vessels and specie jars – large decorative cylindrical lidded containers. In 1775 German physicist Georg Christoph Lichtenberg wrote that Cheapside looked 'as if it were illuminated for some festivity: the apothecaries and druggists display glasses filled with gay-coloured spirits'. Placing carboys in windows may have had a practical use originally for preparing tinctures by maceration, which required heat and light, but they soon developed a purely decorative function. Formulae for coloured liquids with which to fill them were published repeatedly, the earliest in Gray's *Supplement to the Pharmacopoeia* in 1821. Meanwhile specie jars also escaped their prosaic origins as large storage containers, perhaps first

PHYSIC.

Pub.d Oct.r 14 _ 1825 by W. Cole. 10 Newgate S.t

Fig. 181 (opposite)
H. Heath, *Physic*, coloured etching, 1825. The interior of this pharmacy shows the large glass window panes that clearly display the coloured carboys and jars, alongside ceramic pots.
Wellcome Collection.

Fig. 182 (right)
Thomas Rowlandson, etching, c. 1793. In this scene from Tobias Smollett's *The Adventures of Roderick Random* (1748), Mrs Lavement arrives home to her husband's apothecary shop, depicted with drawers, jars, a hanging alligator and specie jars and carboys displayed in the window.
Wellcome Collection.

Fig. 183 (below)
D. T. Egerton, *A Young Man Outside a Pharmacy*, aquatint. Contemporaries looking at this early 19th-century image would have no doubt that the shop was a pharmacy from the window display, despite there being no sign.
Wellcome Collection.

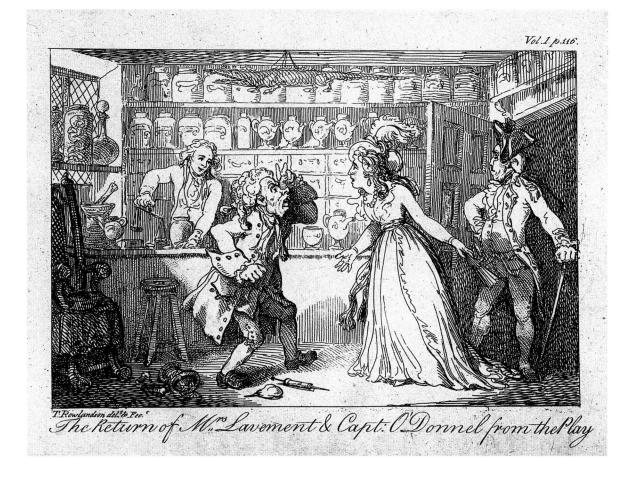

Vol. 1. p. 116.

T. Rowlandson del.t & Fec.t

The Return of Mrs. Lavement & Capt: O'Donnel from the Play

VIS A VIS

Design'd & Etch'd by D. T. Egerton.

It may sometimes be necessary to pay a visit to the Country (Rus in Urbs) in order to avoid the importunities of an unceremonious creditor, should your ill stars throw you plump in his face in turning a corner you may find it no trifling bore; and will curse the inflexibility of your countenance, which spite of your attempts at distortion, will not suffer you to pass without recognition.

simply being stored in the window due to lack of space, but taking on symbolic value gradually. They first emerged in suppliers' catalogues in the 1830s, decorated with gilding and coats of arms, alongside enormous pear-shaped and swan-necked carboys (figs 182 and 183). Both were intended to entice the customer with the promise of quality and entrepreneurial success.

By 1834 the Master of the Apothecaries, its senior officer, reported that only six members were practising pharmacy exclusively. Legal and legislative developments such as the Rose Case of 1704/5 and the Apothecaries Act of 1815 shifted some power from the College of Physicians to the Society of Apothecaries, meaning that apothecaries were able to choose a medical path, leaving the chemists and druggists to take the medicinal one. At the cutting edge, this meant scientific laboratory-based pharmacy, and a search for the active principles or alkaloids of the most popular tropical medicinal plants. European chemists led this search. The initial roll call read: morphine from opium (1804), emetine from ipecachuana (1817/8), strychnine from Strychnos nux-vomica (1819) and quinine from cinchona (1820). *Materia medica*, or the scientific study of crude drugs, became a distinct discipline in medical training from the late 1700s. Thus the scene was set for another explosion in pharmacy products, based largely on these new chemicals.

'THE DRUGGIST'S SHOP, VYING WITH THE GIN PALACE IN ITS TEMPTING DECORATIONS': PROFESSIONALISM AND EXPANSION, 1840s TO 1910

Chemists and druggists in the 1840s were subject to stinging criticism. *Blackwood's Magazine* in July 1841 described 'swarms of chemists, who, without education, qualification, or experience, impudently take upon themselves to prescribe for all manner of ailments'.[19] A Select Committee on Medical Poor Relief in 1844 heard about 'The easy access to the druggist's shop, vying with the gin palace in its tempting decorations',[20] and the *Pharmaceutical Journal* admitted in 1843 that, 'Unfortunately in most country towns not only is every Grocer and Oilman a Druggist but almost every Druggist is a Grocer and Oilman.

Fig. 184
Advertisement for Beasley & Jones's Medical Hall, Leamington Spa. Large plate glass window panes reveal a strikingly minimalist display, with the drug run and storage jars behind the counter visible through the open door.

Wellcome Collection.

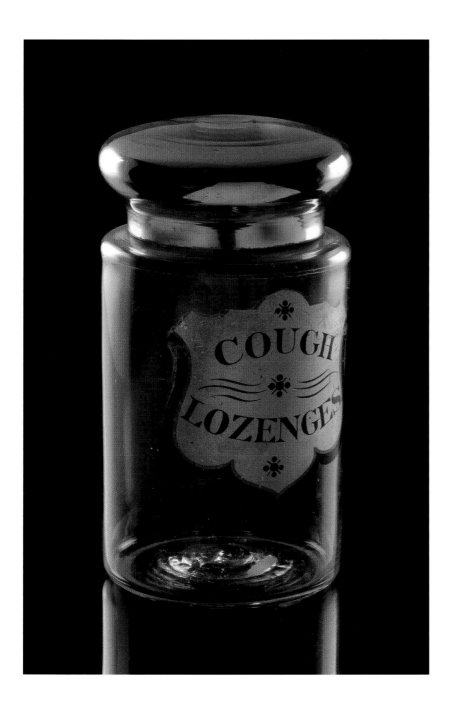

The Druggist has no badge or credentials to designate his superior qualification, in fact, he is not of necessity more qualified than the Grocer.'[21]

In this climate, the Pharmaceutical Society, founded on 15 April 1841, aimed to establish pharmacy as a profession, built on a national membership organisation and a stringent educational scheme leading to an examination-only route to universal registration. They had to overcome the overtones of quackery and the long history of trade, which was viewed superciliously by physicians and politicians. A nationwide survey carried out by the *Medical Times and Gazette* in 1853 found that there was one chemist and druggist for every two medical practitioners. They were rivals: pharmacists provided free advice with the sale of medicines; a general practitioner charged a minimum fee of about five shillings for a visit in the 1860s. Doctors claimed that pharmacists were preying on the poor's ignorance for commercial gain, but the historian S. W. F. Holloway argues that working-class customers valued the chemist's shop as a more dignified alternative to turning to the Poor Law authorities.[22]

As in previous centuries, the pharmacy of the 19th century was recognisable by its window display (fig. 184). Photographs show overwhelming quantities of products, advertising and signage. In 1843 the Hungarian physician J. E. Feldmann wrote:

> When passing along the street, it is a physical impossibility to look into any apothecary's shop, except through the open door, the window is so blocked up with the most multifarious objects. From the middle of the panes glare huge, coloured glasses, yellow, red, and blue, having inscribed upon them certain talismanic characters ... the lower part of the window is occupied, or rather *dressed out*, as the term is, with numerous small bottles for aromatic liquids &c., larger ones with lavender water, bottles of eau de Cologne: horse-hair gloves: syringes of every size and material: an infinite variety of soaps, and, lastly, innumerable boxes of pills.[23]

Inside the Victorian pharmacy, ceramic jars remained a feature, although more uniformly coloured and labelled mass-produced earthenware had taken over from its delftware predecessors. Ceramic jars were still widely used for soft

Fig. 186
Mabel T. Sara, *Apothecary's Shop, Redruth*, oil on canvas, 1903. The brilliant blue of the glass bottles stands out in this painting.

Science Museum Group.
Object number 1989-1172

preparations such as extracts and ointments, but glass was used to store liquids, powders, pills and chemicals (fig. 185). The growing popularity of glass bottles, known as 'shprounds', was matched by technical developments in bottle manufacturing (fig. 186). Having first appeared in the late 1700s, they were a regular sight by the 1840s, with specialised stoppers and neck designs for different substances.

The colours and shapes of bottles became symbolic, as features of delftware jars had been in earlier centuries. The use of specific bottles to denote poisons – usually ribbed and hexagonal for dispensed products, cylindrical within the pharmacy – was not mandatory until 1899, although the 1868 Pharmacy Act required that 'containers for storage [of poisons] should be tied over, capped, locked or otherwise secured in a manner different from that in which other articles are kept.'[24] After 1899, the more established blue poison bottles were matched by a popularity for green glass.

Below the shelves of jars and bottles, the drug run was still commonplace as a practical solution to the raft of prepared and raw ingredients

needed to fulfil customers' demands. In the inventory of Thomas Hughes's Llandeilo shop in 1846 were 100 mahogany drawers, faced and gold labelled, plus 'window globes' and decorative earthenware jars for tamarinds and leeches (fig. 187). Henrietta Cresswell, describing her father's dispensary in Winchmore Hill in the early 1860s, remembered 'an old counter of dark mahogany, shelves of dusty bottles, and a row of drawers with mysterious glittering gold labels, Rad. Quass., Cort. Aurant., and Pet. Nit. etc. If you wanted a cork for a fishing float it must be stolen from "Subera", and in another drawer, labelled "Rad.Zingib." were some pewter "squirts".'[25]

Rather than dispensing medicines on the counter, the trend later in the 19th century was to work out of public view in a small dispensary or behind decorative glass screens. This subdivision of space differed between shops and over time. Robert Drane's shop, opened to his own design in 1868 on Queen Street, Cardiff, had a typical interior with mahogany fittings and glass-fronted cupboards, but there was also a small examination room in the back where Drane would diagnose and treat minor ailments.

Fig. 187
Leech jars, often in a set with honey and tamarind jars, became increasingly ornate in the 19th century. This is a highly ornate example in blue gilt earthenware.

Science Museum Group.
Object number A43107

In Henry Deane's pharmacy in Clapham, London, the 'long dispensing counter ... is well hidden from customers by the upright showcase in the front of it, wherein are displayed brushes and many other toilet-articles.'[26] The skill of the pharmacist thus became concealed from view.

The Victorian customer witnessed an explosion of proprietary medicines. Sales of branded medicines increased nearly ten-fold between 1855 and 1905 (in value from around £0.5m to £4m) while the population barely doubled, reflecting the incredible rise in working-class buying power. This went hand-in-hand with a massive expansion in advertising, particularly for medicinal products. In 1834 the *London Medical Gazette* claimed: 'We can scarcely go into any street in London in which we do not see "Morison's Universal Pills for the cure of every disease" staring at us in large letters in the windows of one or more shops.'[27] By the end of the century, adverts for medicines plastered newspapers, railway stations and trams.

Manufacturers supplied free branded showcases to stock their preparations, whether Pears Soap, Allenbury pastilles or Leath & Ross homeopathic remedies. Stocking animal medicines was also commercially sound, certainly in rural areas, as well as within cities (fig. 188). By producing and selling horse medicines, pharmacies played their part in keeping the country running: in 1902, there were still about three and a half million working horses in Britain.

Pharmacists faced significant competition from stationers, grocers, butchers, hairdressers and publicans, all of whom sold medicines. The number of retail outlets licensed to sell patent medicines increased from 10,000 in 1865 to over 40,000 in 1905, of which only about one-third were run by qualified pharmacists. This increased pressure on pharmacies to sell an expanded range of goods, which exposed a tension with the maintenance of a professional image. The women pharmacists who ran the shop at 17 The Pavement, next to Clapham Common in London, in the early 20th century found that adopting bespoke coats with a sage-green collar, cuffs and belt over their ordinary dresses inspired increased confidence in the customer, who felt that 'they are dealing with women who are specialists in pharmacy (the very thing that customers go into

Fig. 188 (below left) Robert E. Price of Rhyl packed his windows full of stock with signs for surgical appliances, photographic materials, cigars, carbolic tooth powder – and a sheep-shaped board for Cooper's Dip to protect sheep against mites and ticks.

Fig. 189 (below right) Metal shop sign, 19th century. This wide range of products was typical for a pharmacist in the 1800s, including non-medical items, such as ink.

Science Museum Group. Object number A643135

the shop for).'[28] But the massive stock range in many pharmacies perhaps made this professionalism harder to convey. In 1843 the London chronicler Charles Knight explained that simple displays of carboys were being overtaken by 'a most profuse array of knick-knacks, not only such as pertain to "doctors' stuff", but lozenges, perfumery, soda-water powders, &c.' (fig. 189).[29] Thomas Hughes of Llandeilo carried a massive stock list: hair brushes to scrubbing brushes; nail and tooth combs; medicine spoons and egg spoons; dyes and paints. His advertising leaflet also lists biscuits, vermicelli, curry powder, 'powder for blasting' (gunpowder), genuine teas and coffees, cigars and fancy snuff and wax carriage lights.

The historian John Crellin points out that, particularly in small towns, the chemist and druggist was effectively operating as a general store.[30] In Wakefield, in 1853, G. B. Reinhardt was a 'Chymist, Druggist, Tea-Dealer and British Wine Merchant', one of 19 chemists in the town directory, 13 of which were also tea dealers. In 1882 an article titled 'Auxiliary Trades for the Chemist and Druggist' suggested exploiting

avenues related to pharmacy's core business, including spectacles, notably ready-made glasses bought in bulk; and wine, an obvious development from medicinal uses of burgundy as a tonic in damp weather, or port, which was recommended for anaemia. Other areas played on the pharmacists' strengths in chemistry and gadgetry: oils for lighting and machinery; aerated waters, which could be made and bottled on the premises; scientific apparatus, especially if there was a local technical college; and homeopathic remedies.[31] This diversification was commercially prudent, but diluted the sense of a specialist profession (fig. 190).

'SEPARATE THE DRUGS FROM THE DOCTORS': PRESCRIPTIONS AND POLITICAL PRESSURES, 1911 TO 1980s

The National Health Insurance Act of 1911 made a major impact on retail pharmacy. Bread-winners with earnings of less than £160 per annum became eligible for free medical care, including medicines, a significant change from

Fig. 190
Beken and Son, Cowes, Wales, are still trading as photographers, built on their foundations as a pharmacy with a 'photographic depot', c. 1905.

Fig. 191
Allen & Hanburys pharmacy
in London mounted a very
restrained display of the
new drug penicillin in the
late 1940s.

the patchy medical insurance arrangements previously in place through friendly societies and other organisations. After significant political wrangling, pharmacists took over the majority of dispensing, which had previously been carried out in doctors' surgeries. This was in stark contrast to one provincial pharmacy that reported dispensing only 43 prescriptions in the previous nine years. By 1937, virtually all chemists' shops were participating: nearly 15,000 shops, dispensing around 65 million prescriptions. This new primary focus on dispensing drugs, rather than providing advice and selling proprietary medicines, intensified when the National Health Service extended free medical care to all from 1948. The immediate increase in the level of NHS prescriptions is staggering: 187 million in 1948, 250 million in 1956. Most prescriptions needed to be made up individually so dispensaries were expanded and shop areas reduced: pharmacists effectively became invisible in the back room.

The NHS also changed the business model for community pharmacies, who received a practice allowance and dispensing fees for items supplied on prescriptions from the Department of Health.

A letter to the *Chemist and Druggist* in this period[32] stated that a pharmacist could finally live from the proceeds of dispensing and the sale of 'Over The Counter' (OTC) medicines, in marked contrast to earlier decades when balancing the books required a full range of proprietary medicines, toiletries, photography, dentistry and optics. Pharmacies, however, still retained this breadth of stock. A piece in the *British and Colonial Pharmacist* in January 1935 set the tone that continued into the post-war period: 'We must remember that to-day the public is deeply concerned with health and beauty and it is to the pharmacy that people and especially women turn for aid in this direction …'. In 1952 the high street chemist Boots pioneered a new method of allowing customers to select drugs and treatments themselves. Known as 'self-service', it provoked the opposition of the Pharmaceutical Society of Great Britain (PSGB), as they maintained that medicines should not be treated in the same way as other goods. The Court of Appeal found in favour of Boots against PSGB in 1953, and customers were able to self-select their general-sale medicines.

Fig. 192
Society of Apothecaries pharmacy, London, 1935. Traces of the 300-year history of elixirs and potions can be seen in this image, with early carboys lining the top shelf.

Wellcome Collection.

Medicines themselves changed rapidly through the 20th century. Concern about the control of dangerous drugs and biological therapies, particularly in the 1920s and 1930s, introduced a heightened conflict between the ongoing need to encourage customer spending and loyalty, and an increasing role as guardians of public safety. At the beginning of the 20th century, customers commonly bought small quantities of ingredients, or requested a favourite remedy to be made up. Alternatively, 'nostrums' made on site, typically cough mixture or indigestion remedies, were still often more popular than national brands. The 'therapeutic revolution' of the 1950s and 1960s massively increased the breadth of available medicines once more, and also meant that manufacturers supplied them as ready-to-use drugs, mainly in the form of tablets and capsules (fig. 191).[33] Analysis of one London pharmacy's records by the historian Stuart Anderson shows that, in 1900, over 60 per cent of prescriptions were for oral liquids and ten per cent were solid (mainly pills and cachets). By 1980, 70 per cent were solid doses (mainly tablets and capsules) and only seven per cent were liquids (mainly

elixirs and syrups) (see by contrast fig. 192).[34] These developments meant that pharmacists no longer needed to stay out of sight in their dispensaries, yet many continued to do so. By the late 1970s, community pharmacists came under political pressure. Now they were wholly university-educated, the investment of public money into their scientific qualifications was questioned in some quarters.[35] At the British Pharmaceutical Conference in 1981, the Minister of Health, Dr Gerard Vaughan, observed that 'one knew there was a future for hospital pharmacists, one knew there was a future for industrial pharmacists, but one was not sure that one knew the future for the general practice [community] pharmacist.'[36]

'ASK YOUR PHARMACIST': VISIBILITY AND COMPETITION, 1980s TO PRESENT

In 1982 the 'Ask Your Pharmacist' campaign, launched by the National Pharmaceutical Association, first appeared in women's magazines. It aimed to re-establish the high street shop as the

Fig. 193
Wallas and Co., watercolour,
20th century. They
occupied their London
premises for more than
50 years from just after
1900. Their classic window
display with two swan-neck
carboys has been captured
in this image.

first port of call for medical advice and to expand the pharmacist's public health role. This direction of travel was consolidated by the findings of the Nuffield Report, ordered by the Nuffield Foundation trustees in 1986, which resulted from a thorough investigation of all aspects of pharmacy. The comprehensive recommendations for community pharmacy included the introduction of smoking cessation schemes and services to drug users to capitalise on the accessibility of the pharmacy on the high street. These developments brought increased numbers of patients into the shop and helped to politically rehabilitate the pharmacists' reputation. This trend has continued in recent years: the NHS community pharmacy contract, agreed in 2005, introduced the supply of Emergency Hormonal Contraception (EHC) as well as chlamydia and diabetes screening services by pharmacies. Separate consultation rooms in pharmacies have been installed to allow privacy when advising patients with sensitive, complex or chronic conditions via Medicines Use Reviews (MURs). However, most of a pharmacist's income continues to come from dispensing prescriptions and the sale of proprietary medicines.

With echoes from previous centuries, the high-street pharmacy faced competition from alternative retailers – first from the emergence of the supermarket, initially for the sale of non-prescription drugs, and then for prescription medicines from in-store pharmacies. This has subsequently been trumped by the sale of medicines online, with the traditional shop being rendered completely obsolete in this transaction – and a whole raft of ethical and regulatory challenges to accompany the concept and practicalities of internet pharmacy. Today, there is a wide range of outlets for medicines and medicinal advice, although the concept of multiple ways to buy drugs is, of course, nothing new.

As for the relationship between the customer and the pharmacist, with 'original pack' dispensing as the norm, pre-packaged to include patient information leaflets, and computer software that produces labels automatically with directions and warnings, it could be, and has been, claimed that pharmacy shops are no longer needed. Arguably, pharmacists' professional status has been eroded by their promotion as an always-available alternative to taking up the time of the (more highly valued) general practitioner. The growing remit of pharmacy technicians, who frequently dispense medicines before a final check from pharmacists, also alters public perceptions. The current decline in the number of high street branches is more visible to the public than the recent introduction of pharmacists in doctors' surgeries, and the pressure to increase pharmacist involvement in residential homes, as well as their ongoing role in hospitals and industry.

In some ways, community pharmacists are returning to their apothecary roots with more clinical services – a shop that includes consultation areas and treatment for minor ailments – as well as a broad range of products. The decor of modern pharmacies also often harks back to the past to represent the profession's long-standing solidity and trustworthiness, with a clutch of jars and shoprounds placed proudly on clinically white shelving units, or a carboy as a logo. However, a tension exists between that pride in the tradition and a desire to present the profession today as being at the cutting edge of science and retail. In addition, there remains an ongoing tension between community pharmacy as trade *and* profession. Just like the apothecary, the pharmacist has to answer to the twin demands of commercialism and professional altruism. But the value of the shop, of personal interaction with an expert in medicines at the point of purchase, of seeing a range of products in order to make an individual choice, of entering a retail space as a consumer, is one that seems set to endure.

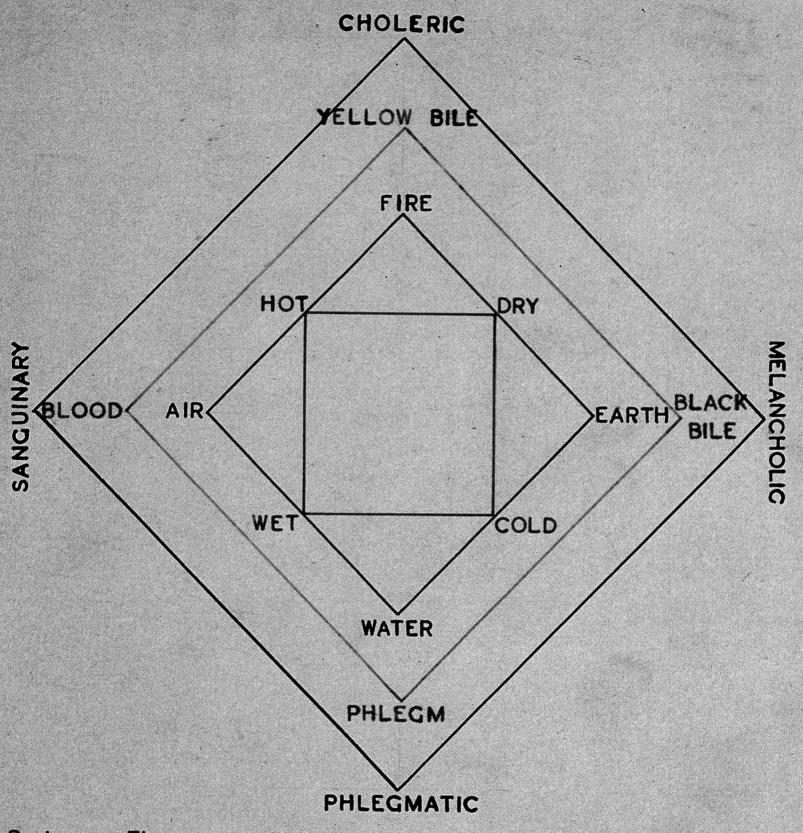

CHOLERIC

YELLOW BILE

FIRE

HOT　　　　　DRY

SANGUINARY　　BLOOD　AIR　　　　　EARTH　BLACK BILE　MELANCHOLIC

WET　　　　　COLD

WATER

PHLEGM

PHLEGMATIC

Red ——— The four qualities　　　　*Sienna* ——— The four humours
Green ——— The four elements　　　*Blue* ——— The four temperament

9
MODELLING LIFE
EXPLORING THE HUMAN BODY

MURIEL BAILLY

Fig. 194
The four qualities,
elements, humours
and temperaments,
c. 1900s.

Wellcome Collection

What are living bodies made of? How does life happen? Since time immemorial, philosophers, physicians, doctors and artists have puzzled over these questions, often building models to test their theories and share their findings. Over the centuries these have taken many shapes, from wax anatomical models exploring the body's 'architecture' to molecular models aiming to reveal its invisible components. The profusion of anatomical and molecular models created over time demonstrates our boundless fascination with the human body and its mysterious workings. Looking at these objects today takes us on a journey, following the transformation of our understanding of the human condition. These models are intrinsically human, not only because they represent the human body (sometimes as a whole, and sometimes a small part in great detail), but because they are made by individuals whose personality and creativity transpires through the models. Such objects are key in allowing us to appreciate humanity's efforts to unravel the body's secrets, and with it the secret of life itself.

In ancient Greece, at a time when dissection of the human body was prohibited by religious beliefs, philosophy and theory prevailed. The Greek physician Galen (AD 130–210) popularised the theory of the four humours, which then dominated Western medicine for centuries to come. This theoretical model described the human body as being composed of four fluids: black bile, yellow bile, blood and phlegm. Each was associated with one of the elements: earth, fire,

air, water respectively. The belief was that an excess of one or more humours in the body would have consequences on personal character as well as health. Someone with too much phlegm would be phlegmatic, someone with too much black bile would be melancholic, someone with too much yellow bile would be choleric, and someone with too much blood would be sanguinary (fig. 194).[1] To avoid such excesses, the role of medicine was to maintain the balance between the four fluids, and treatments consisted in adding or removing the humour in question. This theoretical framework is one of the earliest comprehensive 'models' of the human body's workings.

FROM THEORY TO SCIENTIFIC EVIDENCE

Galen gained most of his knowledge of human anatomy as doctor of gladiators in ancient Rome, mending broken bones and sewing up deep cuts. Because religion and public morale condemned the dissection of cadavers, he was prevented from carrying out a deeper investigation of the body. To fill the gaps in his knowledge, he dissected animals – dogs, pigs and especially monkeys (considered the closest to humans) – and transposed his observations onto humans. Through his dissections and consideration of human anatomy, Galen sought visual and physical evidence of the four humours. Rather than starting with an open mind and drawing conclusions from his observations, Galen had as his point of departure the theory of the four humours, which he then insistently applied to the bodies he dissected. His vision of human anatomy remained unchallenged until the Renaissance (c. 1300–1700), a period characterised by a shift towards rigorous, evidence-based, scientific investigation. Human dissection was still controversial not only because it was close to death but also due to the belief that, when the body is cut open, the soul is disturbed and thus prevented from travelling to the afterlife. Andreas Vesalius (1514–1564) pioneered a new scientific approach and was the first to systematically dissect cadavers in order to explore human anatomy. He published his findings in his 1543 book *De humani corporis fabrica* (*On the Fabric of the Human Body*). His observations, many of which contradicted the Galenic teachings on which medicine was based at this time, were captured in exquisitely detailed drawings (fig. 195). Scientific observation based on human dissection challenged a well-established medical practice upon which many had built their fortune and reputation. Despite the hostile reception of his work from such practitioners, many followed Vesalius's injunction to dissect human bodies and work from scientific evidence. Perhaps one of the most famous of his followers was William Harvey (1578–1657) who, through dissection of the dead and experiments on the living, demonstrated that blood is not produced in the liver, as Galen had stated, but that is pumped around the body by the heart.

The study of anatomy, the opportunity to peer inside the human body and witness its mysterious workings, fascinated medical students and the public alike. As a result, medical schools offering anatomy lessons, many of which were open to the public, flourished all over Europe. The only obstacle to this thirst for knowledge was the lack of primary resources: cadavers. In many of these schools, due to religious beliefs, dissection was only permitted upon executed criminals. There were soon insufficient felons to meet the growing demand for bodies. To overcome this, anatomists and artists collaborated to create wax models and detailed drawings that captured the beauty and complexity of the human body. The incentive to share this new knowledge prompted the proliferation and dissemination of these drawings and models to the many new anatomy schools across the continent.

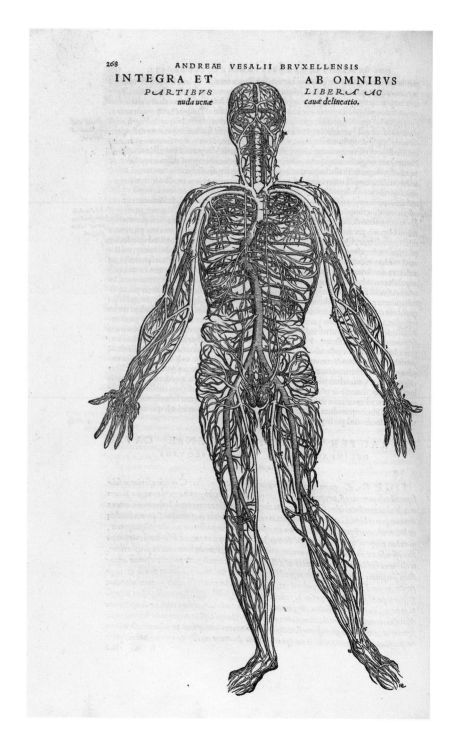

Fig. 195
Andreas Vesalius, veins and arteries from *De humani corporis fabrica*, 1543.

Science Museum Group.
Object number E2010.342.19

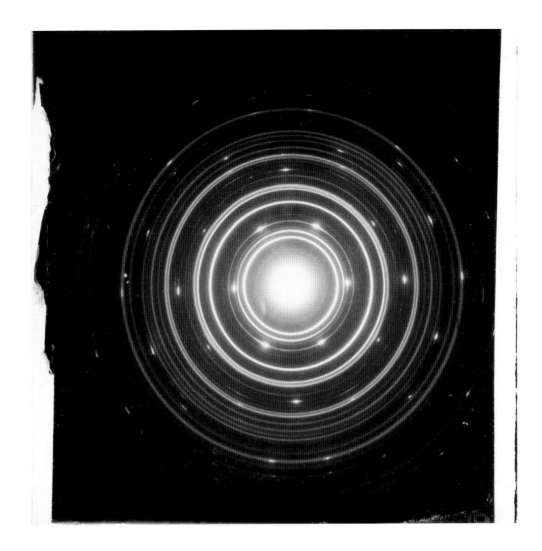

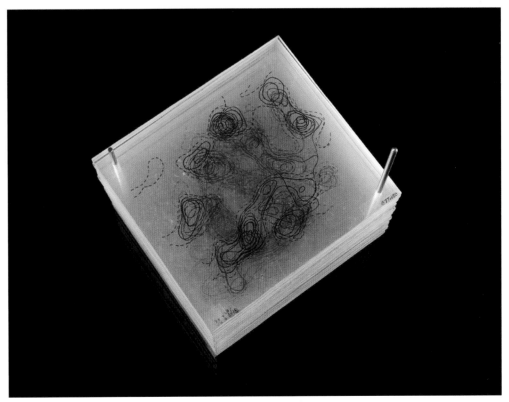

Four hundred years later, X-ray crystallographers were building models with the very same intention of revealing the body's secrets, only with models that had moved from the anatomical to the molecular scale. X-ray crystallography was pioneered in the early 1900s by William Bragg (1862–1942) and his son, Lawrence, on the back of the earlier discovery that 'all matter is composed of a mass of tiny interlocking crystals'.[2] The process consists of freezing molecules at very low temperatures to crystallise them. When X-rays are applied to the crystallised molecules they bounce off the internal structure before reaching the X-ray film. The resulting image shows dots where the rays have hit the film. The intensity of the dots, that is, how light or dark they are, reveals the concentration of electrons in a specific area of the structure (fig. 196). These images are used to create the electron density map, which, in turn, is used to determine the three-dimensional structure of the molecule (fig. 197).

Post-war Britain was characterised by economic growth and a strong enthusiasm for science. Radar technology, early computing, penicillin and the atomic bomb all resulted from scientific research and were considered key contributions to winning the war. As a result, the 1950s saw rising science budgets lead to flourishing research in the field of molecular biology, revealing the structures of the molecules and proteins that make up the human body. Understanding the physical and chemical structure of the body's invisible 'building blocks' was key to understanding life itself. Britain led the way in this field, attracting scientists from all around the world and becoming a hub for international collaborations. The discovery of the three-dimensional structure of deoxyribonucleic acid (better known as DNA) by the Anglo-American duo Francis Crick (1916–2004) and James Watson (b.1928) in 1953 gave rise to what is undoubtedly the molecular model best known by scientists and the public alike. Its familiarity is due both to the iconic model itself and the hugely important scientific and medical consequences of the discovery it represents (fig. 198). Indeed, Crick was credited until very recently with the statement that they had 'found the secret of life' by successfully completing the model. Although Watson has since confessed that he had made this anecdote up to add emphasis to the discovery,[3] the statement certainly has some truth to it.

MODELS: ICONS OF THEIR TIME AND OPEN WINDOWS INTO THEIR MAKER'S MIND

Anatomical and molecular models were both trusted tools for revealing and conveying new knowledge about the human body and its workings. Although born of a purely scientific endeavour to understand the human body, the models are tinted with subjectivity, revealing as much about their maker as they do about the human body. The exquisite anatomical Venuses of Clemente Susini (1754–1814) reveal the artist's meticulous approach and painstaking attention to detail (fig. 199). Susini was the lead wax modeller at the Museum of Physics and Natural History in Florence, known today as La Specola. Founded in 1775, it is one of the oldest scientific museums in Europe. Susini's dissectible, life-size anatomical Venuses were conceived with the dual intention that they would be museum objects on display for the public as well as scientific objects used for teaching anatomy. He produced hundreds of models, many of which can be found today in major museum collections across Europe. The example in the Science Museum's collection is small in size and therefore is considered to be a preparatory work for a full-size version.[4] Despite this, it is complete with several layers of dissection and detailed internal organs, and is adorned with real human hair and Venetian-glass eyes. Susini's models are also characterised by lavish poses and accessories. The level of detail and the romanticised image of the female body showcases Susini's skills as an artist, but it also serves the scientific narrative by distancing the models from the macabre study of death. The perfection of the anatomical Venuses, inside and out, illustrates the values of the time in which 'to know the human body was to know the mind of God'.[5]

Susini's Venuses appear to be in a state of abandon, almost as though they are asleep, fully subjected and oblivious to the dissection process. In comparison, Joseph Towne's models are fraught with life. Towne (1808–1879) worked as wax modeller at Guy's Hospital Medical School in London for 53 years, during which he produced hundreds of anatomical, dermatological and pathological models made from observations on cadavers as well as living patients. Unlike Susini's, Towne's anatomical models are brutally realistic: the 'organs appear

Fig. 198
Crick and Watson DNA molecular model, 1953. Reconstruction of the original double helix model of DNA, using some of the original metal plates.
Science Museum Group. Object number 1977-310

OVERLEAF
Fig. 199
Clemente Susini, anatomical figure of a reclining woman, wax, Florence, 1771–1800. Model with fully removable internal organs.
Science Museum Group. Object number A627043

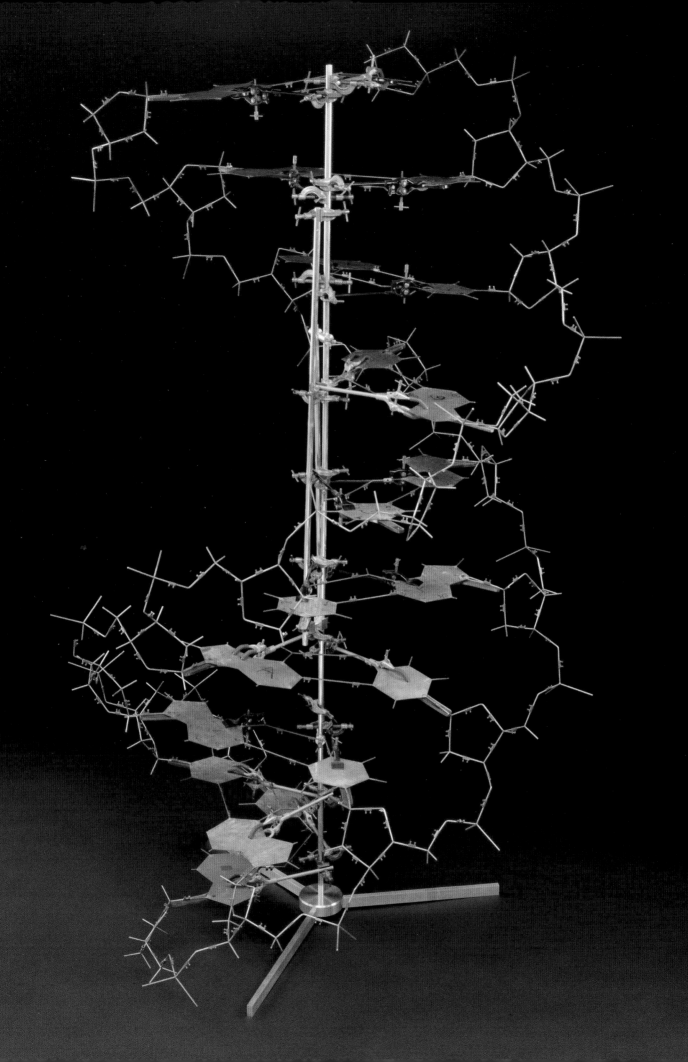

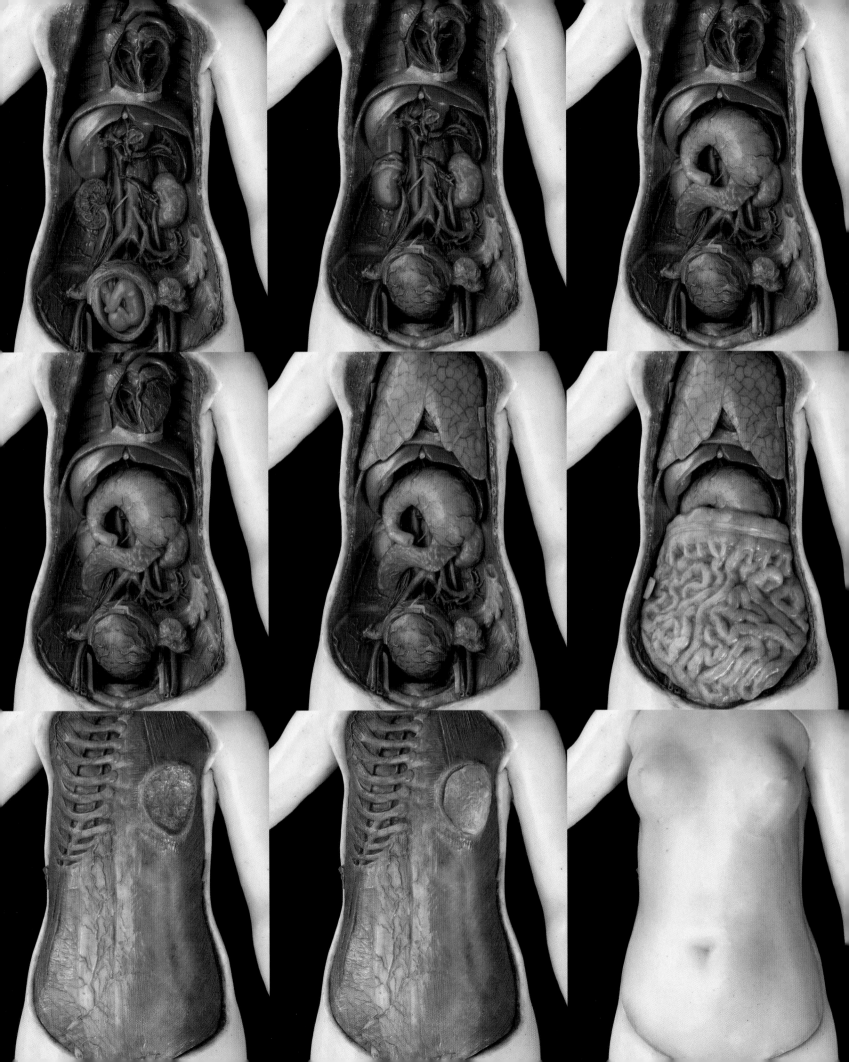

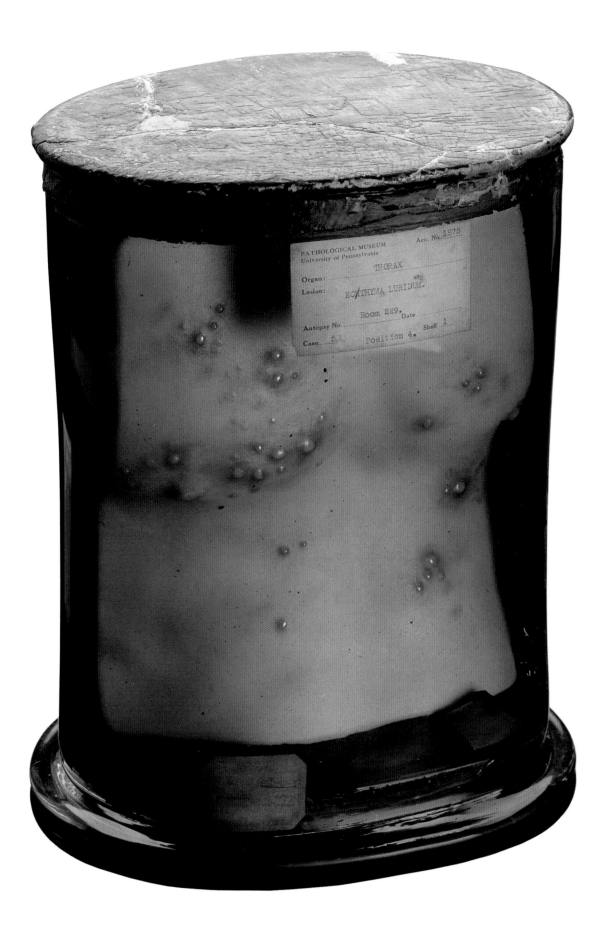

On the model's label:

PATHOLOGICAL MUSEUM
University of Pennsylvania

Acc. No. 1875

Organ: THORAX

Lesion: ECTHYMA LURIDUM.

Room 229.

Autopsy No   Date

Case   5.   Position 4.   Shelf 1

Fig. 200 (left)
Joseph Towne, anatomical
model of a thorax showing
Ecthyma luridium, wax,
London, 1825–79.

Science Museum Group.
Object number 1986-455

Fig. 201 (opposite above)
Joseph Towne, anatomical
model of a hand showing
phagedenic ulcer of the
thumb, wax, London,
1825–79.

Science Museum Group.
Object number 1986-457

Fig. 202 (opposite below)
Joseph Towne, anatomical
model of a torso showing
the exposed internal
organs, wax, London,
1825–79.

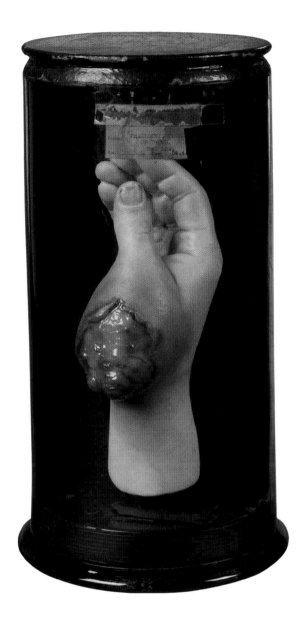

to be inflated with gases, discarded layers of flesh which almost ooze … these works are the anatomist interrupted'.[6] Towne, like Susini, created very detailed models using real hair and eyelashes, but his models represent real people and are not allegories of death. As well as studying the dead, Towne examined the living and their diseases. This sense of individuality is exacerbated in Towne's pathology specimens. These, made from living patients, captured and documented the evolution of skin diseases. They were often *moulages* (casts) taken straight from the patient, thus retaining their unique features (figs 200–2). The abundance of skin disease models produced by Towne represents a shift in the medical and social context from Susini's time. In Renaissance Italy, bodies available for dissection were scarce and mostly limited to executed criminals. As the study of anatomy became increasingly popular, Italy led the way, but anatomy schools flourished all over Europe. The demand for cadavers with which to teach far exceeded the supply available, giving rise to a black market of corpses. Grave-robbers, or body-snatchers, stole freshly buried corpses from graves in order to sell them to anatomists. When bodies could not be obtained from graveyards, some reverted to murder. The most notorious example was that of William Burke and William Hare in Scotland, who killed 16 people over a 12-month period in the early 1800s (see p. 73). The public outcry that followed the scandal

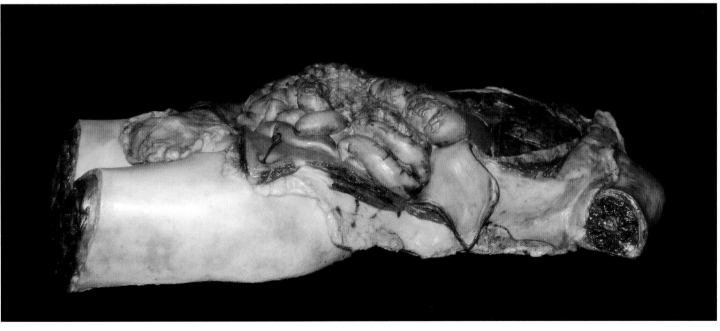

prompted the passage of new legislation in the United Kingdom. The Anatomy Act of 1832 came six years after Towne took his position at Guy's Hospital Medical School. This allowed anatomists to perform dissection on unclaimed bodies. Therefore, for most of his career, Towne did not have to disguise his models as being anything other than derived from cadavers, either to distract his contemporaries from the controversial study of death or to elevate the study of anatomy to an abstract ideal. His contact with patients also gave him access to a more human dimension of medicine, investigating how diseases affect the living body. Towne's high number of pathology models also illustrates the rise of dermatology in the mid-19th century. By displaying his works in jars as if they were real human specimens, he demanded they be thought of as teaching tools as opposed to museum artefacts, thus asserting the scientific and educational purpose of his work.

Molecular models might seem less straightforward than anatomical models at first glance, but in fact they too can tell us a lot about their maker and the context in which they came

to be. Many will be familiar with the Crick and Watson DNA model, as well as with the term 'DNA' itself and the idea that it defines who and what we are. We are increasingly familiar with the concepts of genes and genetic manipulation that are ubiquitous in the media and popular culture. But all these ideas and knowledge are ultimately the consequences of the discovery of DNA's molecular structure. More than just an illustration of the structure itself, the model tells the story of this discovery. Crick and Watson had been trying to solve the three-dimensional structure of DNA for a few years with no success, when in 1952, Maurice Wilkins (1916–2004), a good friend of Crick, who was also working on DNA at King's College in London, showed them an X-ray photograph of DNA taken by his colleagues Raymond Gosling and Rosalind Franklin. This photograph (fig. 203), much reproduced since then as 'Photo 51', gave them the clues they needed to attempt to solve the structure of DNA.

It is this moment of scientific quest, trial and error and questioning that the DNA model embodies. If one looks carefully at the model,

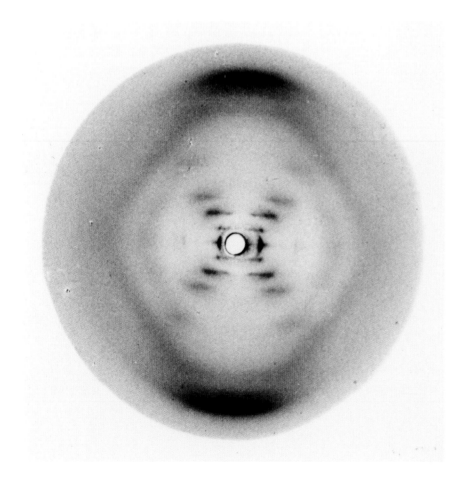

Fig. 203
Rosalind Franklin and Raymond Gosling, 'Photo 51', X-ray diffraction image of crystallised DNA, 1952.

King's College London

the letters 'A, C, T, G' (standing for the four amino acids that constitute DNA) can be seen handwritten onto some of the plates (fig. 204). In some cases, deletions and corrections are also visible, offering the viewer some insight into Crick's and Watson's thought process. These notes which they scribbled on the plates as they assembled their model are all that remain from this pivotal moment in scientific research. The final result, the 'double helix' structure of DNA, has become a symbol of scientific progress. Pictures and diagrams of the structure are reproduced widely in scientific publications and educational textbooks.

The physical model was quickly superseded by the knowledge it unlocked. Crick's and Watson's model was dismantled to be repurposed for further research, while DNA and genetic research was blossoming. It was only in 1976 that some of the original plates were rediscovered in Bristol and acquired by the Science Museum.

The impact of molecular models can be felt far beyond the realm or medicine. For instance, a penicillin model by Dorothy Hodgkin (1910–1994) is linked to intense political and medical

narratives (fig. 205 and detail). She started investigating the structure of penicillin in the early 1940s, in the midst of the Second World War. The conflict had a strong impact on scientific research, and funding was only awarded to projects that directly supported the war effort. Hodgkin, therefore, had put aside her investigation into the structure of insulin (which she eventually completed in 1969 after 35 years of work), to focus her efforts on discovering the molecular structure of penicillin. In 1928 Alexander Fleming (1881–1955) demonstrated that penicillium mould could kill bacteria, but he declared that the substance was 'too unstable to be isolated and used as a drug'.[7] With bacterial infection a major threat to soldiers' lives during conflicts, the investigation of its molecular structure became crucial in the hope that it could be synthesised and mass produced. Hodgkin succeeded in revealing the molecular structure of penicillin in 1945. The model in the Science Museum's collection is a refined version of her first model, also made in 1945. On this later model, she overlaid the three-dimensional structure over the electron-density map (see fig. 197) that she

Fig. 204
Original plates from the Crick and Watson DNA molecular model (see fig. 198).

Science Museum Group. Object number 1977-300

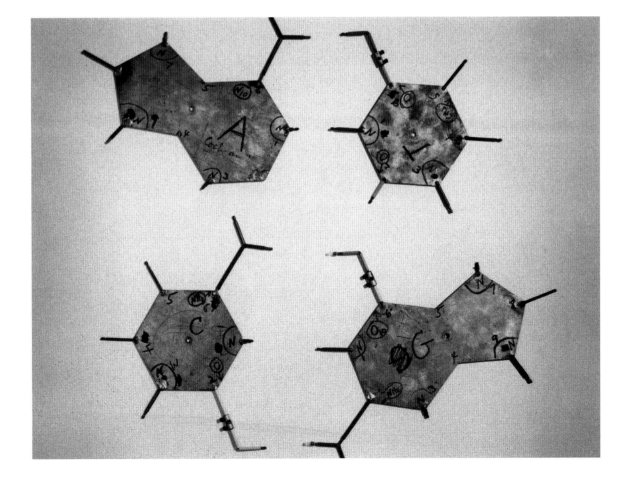

had obtained using the X-ray crystallography technique. The atoms on the model appear as if they are protruding from the map. The superimposition of the three-dimensional structure and the map reveals the scientific process underlying the discovery and proves the accuracy of the model. It is possible that she created such a detailed model to shut down any attack on her structure, which she knew to be correct. Indeed, Robert Robinson (1886–1975), Waynflete Professor of Chemistry at Oxford and a Nobel Prize winner, was heavily involved in the work on the structure of penicillin. Although working alongside each other, Robinson's and Hodgkin's teams took different approaches to the molecule and disagreed on how the three-dimensional structure should look. At the time, penicillin was the largest molecular puzzle that scientists had attempted to solve using X-ray crystallography, and it proved very difficult to investigate due to the instability of the mould and the challenge in growing crystals big enough to be used. Hodgkin's model discredited the theory by Robinson, who had a reputation for not admitting defeat easily. After building her

model, Hodgkin showed it to her colleagues Edward Abraham and Ernst Chain, who apparently joked that she should take a gun with her when showing it to Robinson in anticipation of his fury.[8]

CREATIVITY AND INGENUITY IN REVEALING THE INVISIBLE

The sheer variety of the look and feel of anatomical and molecular models demonstrates the creativity surrounding their making. In both disciplines, model-makers were pioneers, venturing where no one else had ventured before. As dissection became integral to the study of medicine, it revealed new truths about the body's workings. Many models were made in the attempt to capture a glimpse into this defended territory. As well as disseminating new scientific knowledge about the human body, the models made the study of anatomy safer, protecting their users from any contamination or infection that might come from contact with cadavers. The models needed to be accurate and fit for handling. It was by taking

Fig. 205 and detail
Dorothy Hodgkin, Molecular model of penicillin, 1945.

Science Museum Group.
Object number 1996-686

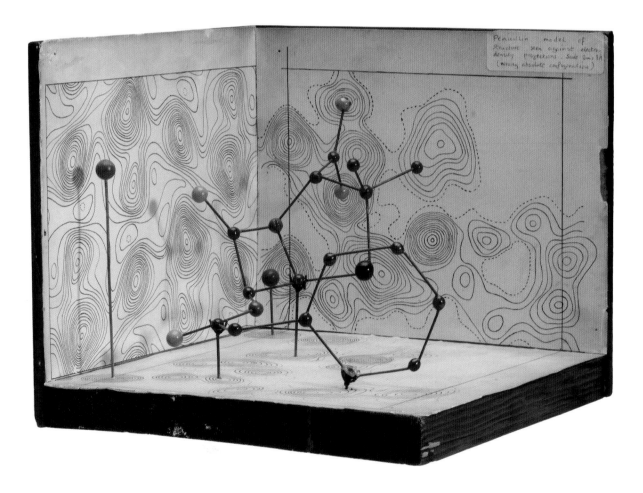

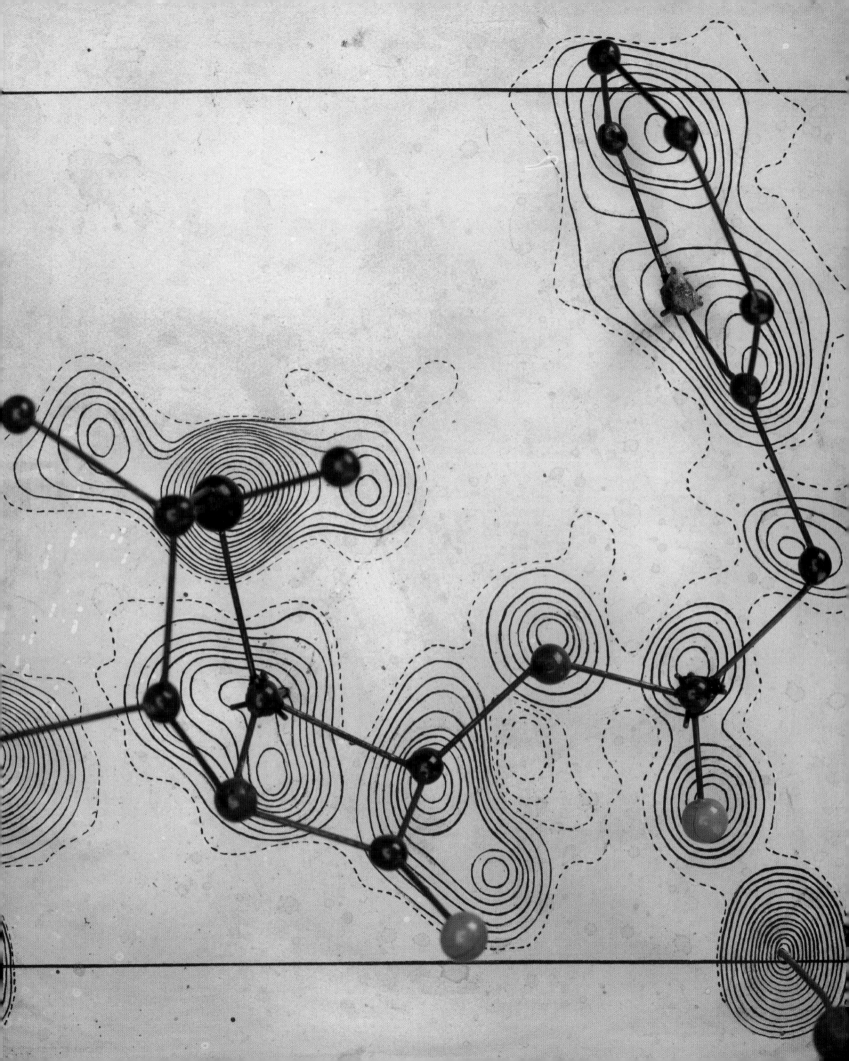

them apart and reassembling them that students truly grasped the structure of the human body. Apart from these two prerequisites, the makers had *carte blanche* on how they could render their subject. Some experimented by injecting real body parts with melted wax,[9] which was then left to solidify. Specimens would subsequently be coated with a varnish. Wax was the preferred medium because of its great malleability and smooth finish, which mimics flesh. But wax is also a very fragile material: if stored at high temperature it sags and melts; if handled too much – and handling was, of course, the models' primary purpose – it cracks and breaks. Louis Auzoux (1797–1880), a French medical student frustrated by the shortage of human bodies to dissect for anatomy lessons, was unconvinced by the adequacy of wax models for this purpose. Auzoux considered them to be too fragile, as well as too time-consuming and expensive to make. He introduced instead papier-mâché models that were sturdier and made up of entirely detachable parts, which he labelled meticulously. They proved to be more suitable for regular handling and teaching than the wax models (fig. 206 and detail).

In the 1950s, the biochemists Max Perutz (1914–2002) and John Kendrew (1917–1997) pioneered the use of X-ray crystallography to study the structure of proteins, the large molecules the body uses to build and maintain itself. They focused their research on blood proteins, Perutz working on the structure of haemoglobin, responsible for carrying oxygen through the blood stream, and Kendrew working on the structure of myoglobin, which stores oxygen in the tissues and muscles, releasing it during effort. In 1957 Kendrew identified the structure of myoglobin – the first ever protein structure to be discovered (fig. 207). After 25 long years of research, Perutz revealed the structure of haemoglobin in 1959. Myoglobin is a smaller and simpler protein than haemoglobin, although it still contains about 2,600 atoms, which explains why it was the first of the two molecules to be solved. To build his first three-dimensional model of myoglobin, Kendrew used plasticine and wooden sticks – probably simply using whatever material was to hand.

The model is a rather crude rendering of the protein. As Kendrew continued to study the structure of myoglobin and to expand his

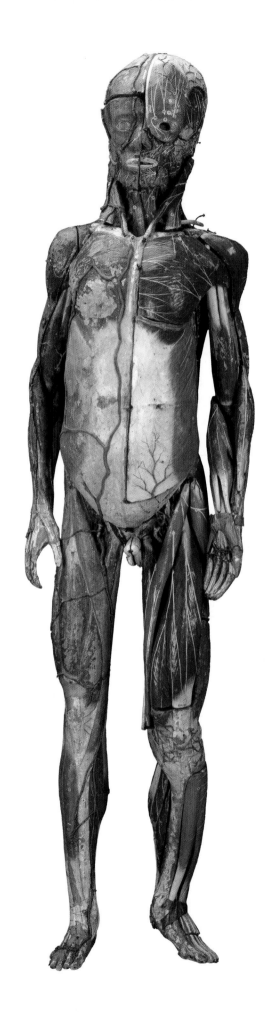

Fig. 206 and detail
Louis Auzoux, anatomical model with muscles, veins and arteries exposed, papier-mâché and wax, France, 1833–66.

Science Museum Group. Object number 1992-707

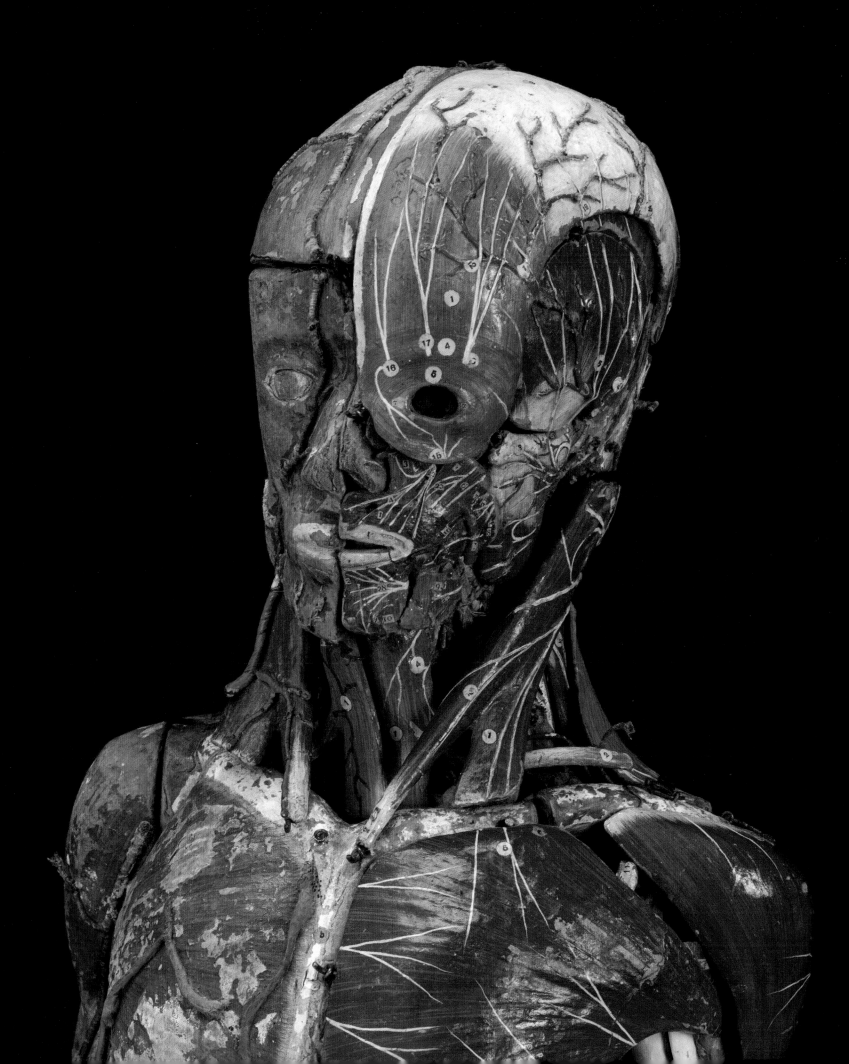

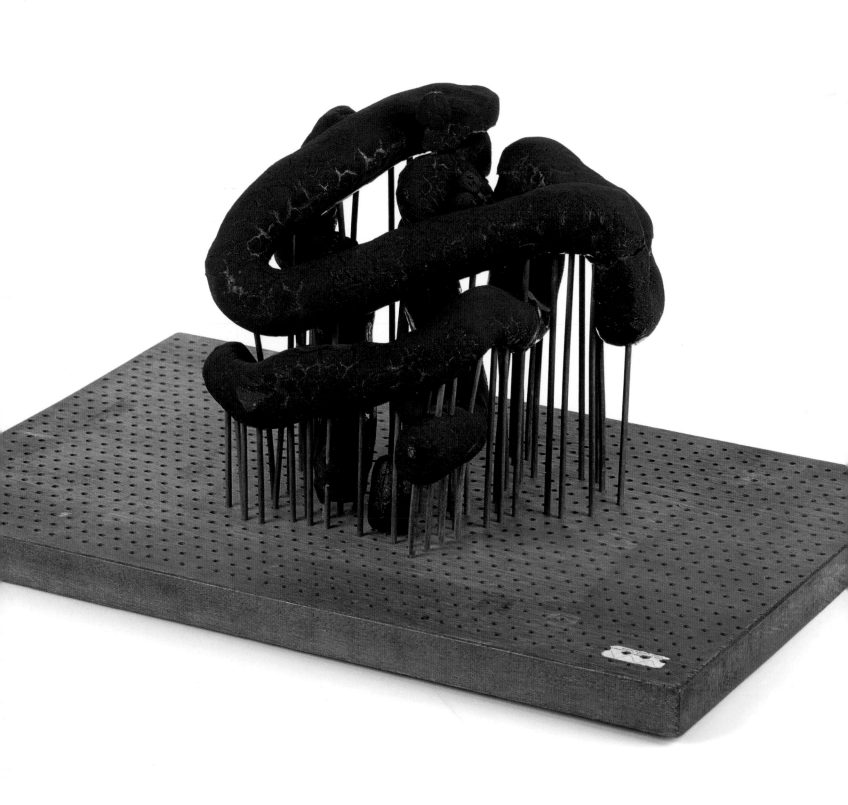

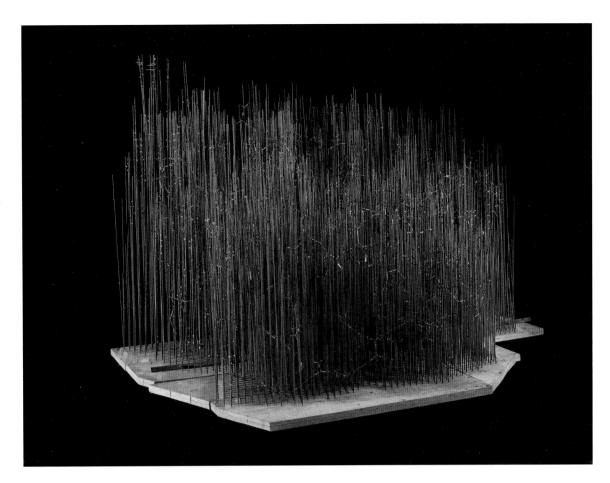

HUMAN HAEMOGLOBIN
1 cm = 2 Å
M.R.C. LABORATORY OF MOLECULAR BIOLOGY
CAMBRIDGE

Fig. 209
Model of horse
haemoglobin, 1967,
copied from the original
by Max Perutz.

Science Museum Group.
Object number 2016-555

understanding of its complexities, he went on to build another, much larger and more detailed version, which is also in the Science Museum's collection (fig. 208). For this second model, he kept the same visual elements, using vertical rods to support the molecular structure. But because of the large scale and higher definition of the protein's structure, plasticine proved unsuitable since it did not allow for enough detail, and collapsed when stretched too far. The first model was already very delicate – and continues to challenge conservators at the Science Museum today – forcing Kendrew to innovate and find new materials to represent his findings visually. He eventually used Meccano toy plastic clips to build his model.

Perutz faced the same challenge from materials when he built his model of haemoglobin (see fig. 209). He too began with plasticine for his first attempt. However, given the complex structure of haemoglobin and the difficulties in working with plasticine, the model collapsed, forcing Perutz to experiment with new materials as Kendrew had done. He eventually resorted to sheets of thermostatic plastic, usually used

to produce electron-density maps, out of which he cut disc shapes that he then assembled together.[10]

For both anatomical and molecular models, the variety of shapes and materials was limited only by accuracy, suitability, 'readability' and the maker's imagination. Where Auzoux carefully labelled all the components of his anatomical models (figs 210–12), X-ray crystallographers often reverted to colour to 'legend' their models. In his large-scale myoglobin model, Kendrew used colours to highlight the electron density of the molecule, while Crowfoot Hodgkin used colour-coding to identify the different atoms composing penicillin.

MODELS' LEGACY AND IMPACT

The models discussed in this chapter represent efforts throughout history to understand the human body and the mystery of life. Using contemporary means and embedded within very specific contexts, they have helped scientists and anatomists to achieve and communicate this understanding. Anatomical models revealed

Fig. 210 (and 211, 212
overleaf)
Louis Auzoux, model of a
human eye, France, 1870
(three views).

Science Museum Group.
Object number 1996-277/11

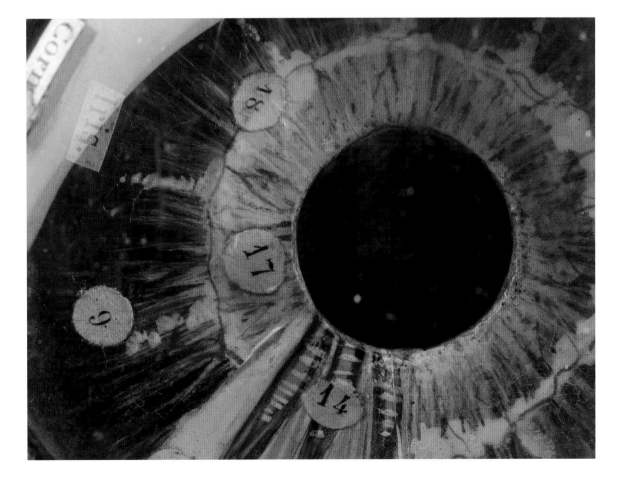

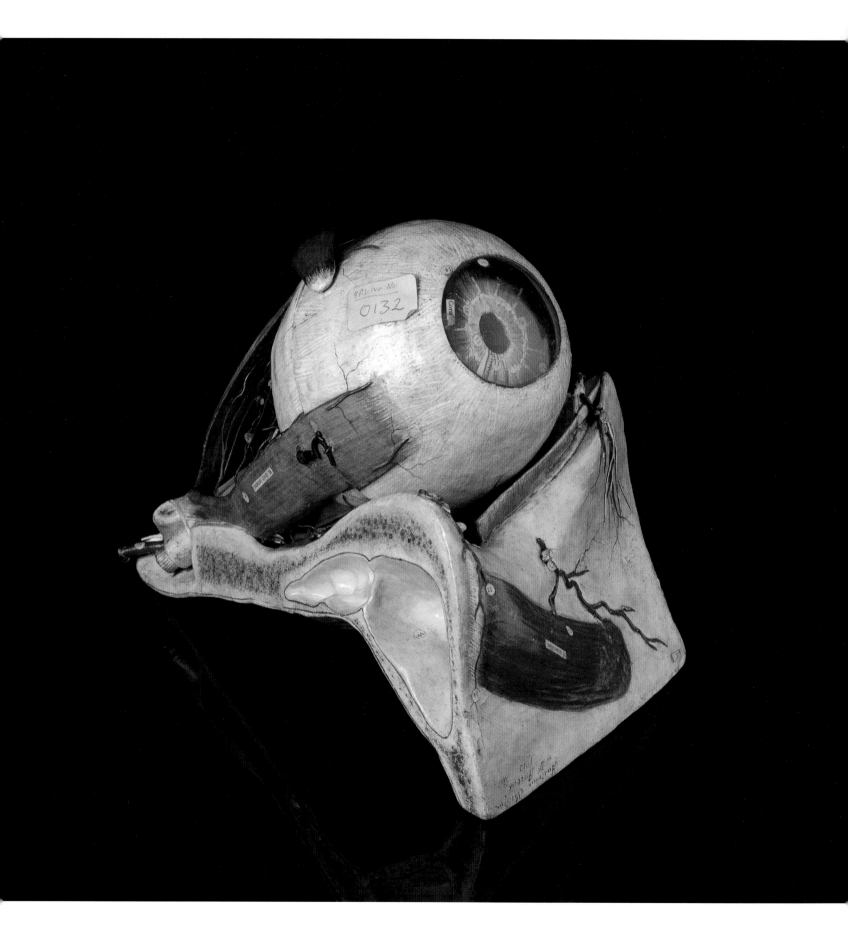

what the human body looks like and how it is organised on the inside, while molecular models show how it functions at the deepest level. The discoveries and knowledge conveyed by these models have had, and continue to have, a tremendous impact on our society, especially in medicine. Dissection, the cornerstone of medical training, led to a scientific understanding of the body. The wide dissemination of anatomical models helped a new generation of doctors to understand the 'mechanics' of the human body. Knowing how organs work and relate to each other, and how they can be affected by diseases, transformed medical diagnosis and treatment. X-ray crystallography provided an ever-deeper view inside the human body. Knowing the physical and chemical structure of molecules allows us to understand how they are supposed to work and consequently enables us to identify malfunctions and thereby design more effective treatments. The discovery of the molecular structure of DNA in 1953 revolutionised medicine, leading to the Human Genome Project, which successfully sequenced the entire human DNA in 2003, and saw the advent of targeted and

tailored gene therapies. Increasingly, gene editing has huge potential to eradicate diseases and cancers caused by genetic mutations. Genetic modifications can be passed on to future generations, thus inscribing any changes into an entire family line as opposed to an isolated individual. Fears about 'playing God', genetic mutation going wrong or the technology being used for ignoble causes, such as engineering humans, are palpable amongst both experts and lay people. Indeed, it seems that there are always two sides to major discoveries: the discovery of the structure of penicillin by Hodgkin was essential to the later development of antibiotics. However, the overuse of antibiotics has led to the rise of antibiotic resistance and, subsequently, superbugs – one of the biggest health threats of the early 21st century.

These frightening concepts and catastrophic scenarios have been adopted widely in movies and literature. Numerous blockbuster films such as *Jurassic Park* and the *X-Men* series (as well as many other superhero offerings) have their origins in dubious genetic research or genetic mutation, while the Nobel Prize Winner Kazuo

Fig. 213
Gunther von Hagens'
Body Worlds exhibition.

Ishiguro's bestselling novel *Never Let Me Go* explores the dark possibilities of cloning. *Oryx and Crake* by Margaret Atwood and the graphic novel *Y, The Last Man* also address these fears in literary form and have attracted the enthusiastic interest of the public.

Even today, anatomical and molecular models continue to pique our fascination and insatiable curiosity about the human body. Testament to this in recent decades has been the success of Gunther von Hagens' ongoing *Body Worlds* exhibition (fig. 213). Exhibits are made using a technique patented by Von Hagens and known as 'plastination', similar to the one described earlier in which molten wax was injected into human cadavers. Von Hagens preserves body parts as well as full bodies by injecting them with a liquid polymer, which is then left to harden. The practice has attracted many fascinated visitors, though the theatricality of the poses in which the bodies are displayed has received much criticism for being sensational and undignified.

Defending himself against critics, Von Hagens argues that his specimens are created as educational tools: 'My work continues the scientific tradition whose recurring theme is that research should serve the general enlightenment'.[11] While the value and meaning of his work is debated, the public interest is clear. 400 years after the first anatomical models were made and 20 years since the creation of the *Body Worlds* exhibition, the show has travelled worldwide and received over 47 million visitors.[12]

During the Renaissance, anatomical models were often made by artists under the close supervision of anatomists to ensure accuracy. Artists practising today continue to look at the human body and its strengths and weaknesses for inspiration. High-profile British artists such as Antony Gormley and Marc Quinn have regularly used the human body, often their own, in their practice. Gormley's work explores how the human body interacts with its environment while Quinn questions body transformation and preservation. In his 'Body Alteration' series, he takes as his subject people who have dramatically modified their physical appearance, 'shaping and sculpting their own flesh'.[13]

Like their anatomical counterparts, molecular models and X-ray crystallography patterns have inspired artists and designers for decades. In 1951,

while the field of X-ray crystallography was prospering and many major Nobel-Prize-winning research projects were under way, molecular structures became a central source of inspiration for the Festival of Britain. Conceived partly as a commemoration of the Great Exhibition that had taken place a century earlier, the Festival aimed to celebrate British industry, art and science. Dr Helen Megaw (1907–2002), an X-ray crystallographer at Birkbeck College in London, initiated the relationship between scientists and designers by contacting Marcus Brumwell (1901–1983), who was at the time the Director of the Design Research Unit. She suggested to him that designers should look at the patterns obtained through X-ray crystallography for inspiration. This collaboration culminated in the Festival Pattern Group producing over 80 designs inspired by the molecular structure of myoglobin, haemoglobin and penicillin, to name but a few (figs 214–16). While retaining their scientific meaning and accuracy, the patterns were turned into fabric and furniture designs printed on synthetic products such as Dunlop's PVC sheeting, and Warerite's plastic laminates. Although they might at first only evoke the nostalgia of a 1950s interior, these molecular structures, as well as the materials they were printed on, embody the boiling creativity and scientific expertise of Britain in the post-war period.

The Festival of Britain, which welcomed 8.5 million visitors, was reported on in *The Times*: 'It is a pleasant change to dissociate the atom from the idea of a destructive bomb and to apply it to the creation of things of beauty.'[14] Indeed, molecular structures still hold a place in the popular mindset. The celebration of 'geek culture' generates countless pieces of merchandise featuring molecular structures, while the popular TV show *The Big Bang Theory* uses the molecular structure of the atom in its title and the DNA double-helix model occupies a central place in the protagonist's living room.

Our fascination with the human body seems to grow alongside our understanding of it. New disciplines offer new lenses through which we can explore and try to make sense of the body and, by association, life itself. The formal scientific investigation of the human body through dissection, from the 1500s onwards, enabled us to better understand the body's workings and how diseases affect us. Five centuries later, dissection is still at the heart of medical training, and wax models as well as pathological specimens are still used for teaching. X-ray crystallography took our understanding of the body and of how life originates and reproduces itself to an unprecedented level. Unravelling the body's invisible 'building blocks' has captured the public imagination just as much as anatomical models had done. With each discovery about the human body and the origin of life come new questions and hypotheses to be investigated. For over 50 years, the discovery of the structure of DNA triggered new research about the potential applications of this new knowledge. In the 1990s Dolly the sheep (the first cloned mammal) embodied the hopes for and fears about genetic research. In the early 21st century this seems like a distant memory as the world explores gene therapy and gene editing with the same range of feelings. Like the making of the models, the application of the knowledge they convey is limited only by our own imagination, which will decide where we will focus our efforts. As Albert Einstein (1879–1955) said, 'imagination is more important than knowledge'.

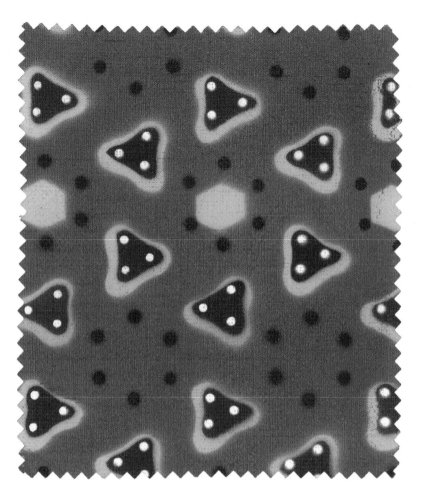

Fig. 214
'Rexine' fabric sample inspired by the basic structure of insulin, Festival Pattern Group, 1951.

Science Museum Group.
Object number 1976-644/1

Fig. 215
Wallpaper sample inspired by the structure of insulin, Festival Pattern Group, 1951.

Science Museum Group.
Object number 1976-644/50

Fig. 216 (opposite)
'Haemoglobin' fabric sample based on X-ray patterns of haemoglobin, Festival Pattern Group, 1951.

Science Museum Group.
Object number 1976-644/7

10 EMOTIONAL OBJECTS
FAITH AND FEELING IN THE MEDICINE COLLECTION

SARAH BOND

SARAH BOND

Fig. 217
Amulets in the stores of
the Science Museum.

Medical collections in general – and the Science Museum's in particular – are fertile sources of emotional objects: artefacts imbued with powerful feelings and beliefs triggered by our confrontations with illness and mortality. This chapter considers three types: votive offerings presented to the gods in the hope of a cure or to express gratitude for healing; amulets and charms, carried or worn to protect against illness or accident or to bring good luck; and modern medical relics capable of producing affective and embodied responses in their viewers.

Despite originating from diverse cultures and time periods, each conveys an emotional charge or resonance that, in turn, fosters an empathic connection to lives and bodies often long since departed. How is this possible, given the temporal, cultural and geographical distances between us – today's museum visitors and curators – and the feelings bound up in these objects' creation and use? Can emotion transcend narrative? And what role might emotional objects play in a 21st-century gallery in a national museum of science?

FAITH

The Science Museum's Medicine collection is unique in many senses. One of the largest and richest of its type in the world, it is composed in large part of the extraordinary private collection of Sir Henry Wellcome. Although most of his collecting actually took place in the 1920s and 1930s, Wellcome subscribed to the so-called 'comparative method' advocated by late Victorian evolutionary anthropologists. According to this scheme, huge quantities of similar items were amassed and arranged chronologically in an effort to establish the origins of cultural practices. Taking his inspiration from the illustrious ethnographical collections of Augustus Pitt Rivers (1827–1900) and Frederick Horniman (1835–1906), Wellcome's goal was to create an all-encompassing 'Museum of Man' that would, in his own words, 'bring together a collection of historical objects illustrating the development of the art and science of healing throughout the ages'.[1]

Wellcome was the first major collector to attempt such a task, and his vast wealth ensured that his plans were not constrained financially. By the time he died in 1936, Wellcome had accrued over a million objects – a remarkable personal legacy tracing the many different ways that people have made sense of, treated and coped with illness over time and across cultures.

As a result of his adherence to a broad, anthropological definition of medicine as anything having to do with 'the preservation of life and health',[2] Wellcome's collection speaks to the human experience of medicine alongside its technology and practice. In the depths of the Science Museum's small-object store, serried rows of obstetrical forceps, stethoscopes and pharmacy-ware jostle with statues of saints and healing deities, ceremonial masks, ancestral carvings, and cabinets bursting with amulets and charms. Unusually for a collection held by a national museum of science and industry, it reflects diverse global healing practices, ranging from masquerade and divination to prayer and pilgrimage – objects relating as much (or more) to faith as to science.

Such artefacts may, at first glance, appear to have little relevance to medicine in the 21st century, particularly as the majority are over a hundred years old. Many of us today enjoy greater choice and access to quality healthcare than at any other time in human history. Where spiritual practices were once a first port of call, we are now more likely to consult a doctor, pharmacist or the internet for medical advice and treatment. Yet while scientific innovation may have transformed our expectations of what medicine can do for us, it has done little to marginalise the powerful role of faith in our experiences of illness and recovery.

In its most inclusive sense, 'faith' refers to the placing of trust and confidence in something or someone, either physical or metaphysical. Depending on our cultural background and the situation, this could be a deity, healer, medical system, technology or product, loved one or community – or, in many instances, a combination of these. Tangible things can also serve as repositories for hopes and beliefs, irrespective of whether it is a pacemaker, paracetamol tablet or lucky pair of socks.

Today, as in the past, our social, cultural and spiritual values influence the decisions we – and those treating us – make in relation to our health. Even when our choices are informed by the best available evidence, we tend to trust the advice we are given and expect the prescribed course of treatment to work – and this confidence itself may in fact contribute to our recovery. Numerous studies have demonstrated that trust plays a critical role in the relationship between patient and caregiver, often resulting in better outcomes. The influential role of belief has been well documented in research investigating the placebo effect (or 'meaning response', as some have more broadly defined it).[3] Faith addresses the fundamentally human need for peace of mind in circumstances that may otherwise seem daunting or insurmountable.

FEELING

When the Wellcome collection was transferred on long-term loan to the Science Museum in 1976, it effectively doubled the size of the collections overnight. Although it had been extensively rationalised in the 50 years since Wellcome's death, it still comprised some 114,000 objects. Science Museum curators have subsequently added over 30,000 additional items to the mix, reflecting ongoing developments in medical practice and culture. So, what is it about this collection that gives it a particular potency and power to elicit emotional responses?

One significant factor is the prevalence of objects serving as literal or metaphorical extensions of the human body. Many – like the aforementioned votives, and the prosthetic limbs described in Chapter 2 – assume the body's forms, while others have been shaped or worn by its touch, or bear the imprints of now-absent bodies. Like most medical collections, this one also contains artefacts that were formerly part of the body itself. Indeed, as the historian Ruth Richardson has eloquently put it, 'in the context of ... Wellcome's collecting, it is ... difficult not to perceive every object as in some sense representing remains of humanity'.[4] By virtue of their acquisition into a museum collection, almost all will outlive the bodies of their creators and users – a fact that only adds to their resonance.

Other items relate to emotionally significant life stages or events associated with powerful feelings and/or bodily changes, such as illness, accident, puberty, pregnancy and birth, war, ageing and death. These transformations can be frightening: they are often accompanied by pain and other distressing symptoms, may threaten our sense of control over our minds and bodies, and ultimately remind us of our mortal status. Not knowing or understanding why our bodies appear to be rebelling is itself unsettling, so we try to identify the causes of our symptoms and ascertain what we can do to make ourselves feel better. Seeking professional medical advice and treatment may well provide answers and (if we are fortunate) solutions, but this is rarely the only place we look to for guidance and support in trying times. Museums, of course, deal in material culture, and so it is perhaps not surprising that the Science Museum's Medicine collection is full of objects that once provided tangible loci for feelings and memories that were difficult to express or articulate. Giving physical form to our emotions can make them easier to manage, particularly in situations where we feel powerless.

The sheer volume of material Henry Wellcome accumulated defies the imagination. He was a collector of collections, habitually acquiring objects by the crateload – many of which were never even unpacked in his lifetime. Like other collectors of his era, Wellcome prioritised an object's position within a series over its own history and collecting took precedence over cataloguing. Consequently, the majority of his purchases have long since been separated from the specific concerns and desires of their makers and users, despite having once been invested with great feeling. Today, their emotional resonance stems as much – if not more – from their collective presence as it does from their individual narratives.

ACTS OF FAITH

One category of objects that echo our bodies' forms are votive offerings in the shape of limbs and organs, unearthed near ancient Greek and Roman sacred sites, where they were once left as gifts for the gods. There are more than 800 of them in the collection, ranging from skilfully cast bronze miniatures to life-sized (and strikingly lifelike) arms, legs and feet, rendered in terracotta. Known as 'anatomical votives', they were mass produced using moulds – typically from local clay – and offered for sale outside pagan shrines and temples. Archaeologists believe that they were purchased and subsequently deposited by pilgrims to these sites to express gratitude for recovery from an illness, or to indicate which part of the body required divine intervention in order to be healed. Although it originated in earlier cultures, the phenomenon thrived in ancient Roman Italy between 400 and 100BC.

While few of them exhibit visible signs of disease, almost every conceivable appendage is represented, from toes, tongues and tracheas (figs 218 and 219) to anatomically ambiguous internal structures – human dissection was taboo in Roman society, so anatomical knowledge was gleaned largely from animal carcasses – flowing locks of hair (fig. 221) and even a tiny bronze pair of what are reputedly tears (fig. 220).

Fig. 218
Votive tongue and tonsils, terracotta, Roman, 400BC–AD200.

Science Museum Group.
Object number A634928

Fig. 219
Votive trachea, terracotta,
Roman, 200BC–AD200.

Science Museum Group.
Object number A636200

Fig. 220
Two votive tears, bronze,
Roman, 200BC–AD400.

Science Museum Group.
Object number A656201

Fig. 221
Two sets of votive hair,
painted black and white,
terracotta, probably
Roman, 200BC–AD200.

Science Museum Group.
Object numbers A634932
and A114891

These are some of the oldest objects in the collection and it is virtually impossible to determine exactly why – or by whom – they were dedicated. The vast majority were found buried in nearby pits, suggesting they were routinely cleared from sanctuaries in order to create room for fresh offerings. The votives themselves, however, do offer some clues as to how they might have been displayed originally. Many have flat bases, enabling them to be placed, freestanding, on the floor or gathered on shelves or altars, rather as they are now laid out in storage (fig. 222). Those that are unable to self-support were probably leant against walls. Others are pierced with suspension holes, indicating that they may once have been hung from a nail or cord, not unlike these examples displayed in the Archaeological Museum of Ancient Corinth (fig. 223).

Theories abound among classicists and archaeologists about the specific motives behind their deposition and what this might tell us about Greco-Roman health concerns and attitudes to illness. While most agree that representing the body in parts was an effective means of drawing attention to the afflicted area, others have pointed out that corporeal fragmentation is a frequently cited metaphor for the experience of being unwell or in pain.[5] Severe headaches, for example, are often described or visualised as 'splitting'.

Certain categories of votive, too, are likely to have served more than one purpose, conveying symbolic as well as literal meanings. Those shaped like eyes and ears may well have indicated problems with sight and hearing, but could also have been intended as a plea for the deity in question to see the dedicant's suffering and hear their prayers (figs 224 and 225).[6] Likewise, legs and feet – which are especially common – may have been offered in gratitude for the successful completion of a long journey (or indeed the pilgrimage itself), rather than in relation to a specific injury or disorder (fig. 226). Interestingly, the relative frequency of found votive body parts varies between archaeological sites, implying that some were specialised centres associated with the healing of particular diseases.[7] Proportionately more eyes were uncovered at the Asklepieion (sanctuary of the healing god Asklepios) in Athens, for example, suggesting that pilgrims

Fig. 222 (opposite)
Votive heads in the stores
of the Science Museum.

Fig. 223 (right)
Votive offerings found near
the Asklepieion at Corinth.

may have travelled here specifically to seek healing from sight-related disorders.

One of the most intriguing types are the heavily ridged, flask-shaped organs that are thought to represent wombs (fig. 227). All have a 'mouth' or opening at one end; some feature an additional pouch-like appendage to the side, possibly denoting the bladder. They vary wildly in form (one researcher identified 48 different types)[8] and bear little resemblance to their real-life counterparts, having perhaps been copied to obscurity from an ancient medical illustration. Some archaeologists interpret the regular bands that decorate their surfaces as being indicative of contractions, or the organ's ability to expand substantially during pregnancy. A handful contain a tiny clay ball that rattles when shaken, perhaps signifying the presence of an embryo. When I first handled these objects in storage, I was struck by the unexpected sound – maybe heard only a dozen or so times in the past 1,000 years. Like other details that emerge only under close observation, it seemed momentarily to bridge the gap between ancient times and the here and now. While we can only imagine the

individual stories that inspired these objects' dedication, it would seem reasonable to assume they were gifted in recognition of a healthy birth, or were a means of seeking reassurance during the risky and uncertain business of conceiving, carrying and delivering a child.

Anatomical votives are vestigial, functioning only as traces of past human experience. They are fragmentary not only in form but also in nature: the offering is all that remains of the encounter, and, in many instances, even the archaeological context has been lost. They are truly anonymous artefacts, separated from their individual narratives – be they of hope, pain, suffering, loss or overwhelming gratitude – in the very act of deposition. In this moment, they undergo a shift in status from private petition to public testament to the divine healing power associated with that site. As just one offering among many, their emotional resonance is amplified by the accumulated hopes and fears of the wider community.

Evidence of votive-giving exists in almost all cultures and it persists today in many faiths, including Christianity, Hinduism and Buddhism.

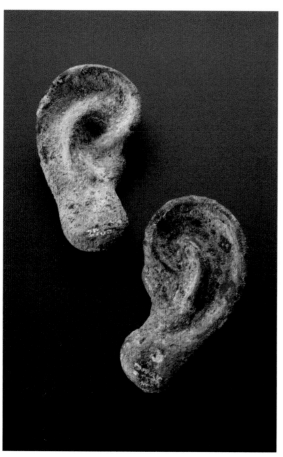

Fig. 224 (above)
Votive eyes, bronze, Roman,
200BC–AD100.

Science Museum Group.
Object number A634942

Fig. 225 (far left)
Pair of votive ears, bronze,
Roman, 200BC–AD100.

Science Museum Group.
Object numbers A634919
and A634920

Fig. 226 (left)
Group of votive feet,
terracotta, Roman,
200BC–AD200.

Science Museum Group.
Object numbers A635655,
A635663, A635656, A635676,
A635669 and A635661

The anatomical variety can still be found adorning altars and saints' statues in southern European churches and across the Catholic diaspora, from Mexico (where they are known as *milagros*) to India – although they are now more commonly made of sheet metal or wax (fig. 228). Their intended function is much the same as that of their ancient equivalents: to express gratitude for healing and/or to commemorate an answered prayer. Surprisingly, Wellcome does not appear to have acquired any non-classical examples, with the exception of two-dimensional *ex voto* paintings,[9] and the area of the collection to which they belong was not actively developed in the years following his death.

This was something I hoped to rectify on a recent collecting trip to the Sanctuary of Fátima in Portugal, a well-known site of pilgrimage constructed around the area where three shepherd children reported witnessing Marian apparitions in the spring of 1917. The small town itself is crowded with souvenir shops and stalls selling – alongside more typical devotional items such as rosary beads, bottles of holy water and figures of Mary – model body parts, figurines and replicas of other prized possessions such as houses, cars and even pets, fashioned out of hollow wax (fig. 230). Available in a range of sizes from miniature to life-sized, these lightly perfumed effigies, known locally as *promessas de cera* (wax promises), are rather more ephemeral than their Greco-Roman predecessors. Once purchased, they are intended to be cast immediately into blazing furnaces in the sanctuary's main square in order that their delicate smoke can drift directly up to Our Lady in heaven (fig. 229). The melted wax is then recycled.

Fátima is the exception rather than the rule, however. In almost every other context, the gradual accumulation of votives in a sacred space over time was – and is – a key part of their affective pull. Seeing them gathered together in vast quantities must have been a powerful and humbling experience (fig. 231). Pilgrims would, at the very least, come away in the knowledge that they were far from alone in their suffering, and the cumulative presence of these objects testified to others' eventual recoveries.

Fig. 227
Group of votive uteri, terracotta, Roman, 200BC–AD200.

Science Museum Group. Object numbers A636083, A636082, A636075, A155134

Fig. 228 (opposite)
Silver-plated *tamata*;
metal votive plaques used
in the Eastern Orthodox
Churches, Greece.

Fig. 229 (right)
Votive candles and wax
offerings being burnt in
the Sanctuary of Fátima.

Fig. 230 (right)
Promessas de cera for
sale at the sanctuary of
Fátima, Portugal.

Fig. 231 (far right)
Votive offerings take
a wide variety of forms
at the Church of Nosso
Senhor do Bonfim, Brazil.

OVERLEAF
Fig. 232
Group of amulets from
diverse cultures and
time periods.

They recall, for me, hospital noticeboards crowded with thank-you cards and photographs from grateful patients and their families, which must also comfort and reassure those still in the midst of treatment or facing an uncertain future. Like the votives, these secular expressions of gratitude mark a conscious decision to publicise the personal, and a subsequent shift in the observer's emotional connection from the individual narrative to the overwhelming mass.

But are anatomical votives capable of evoking emotional responses in the secularised space of the (science) museum? While we cannot assume that all visitors will automatically grasp their significance in the way that pilgrims and worshippers might, with sensitive interpretation we hope they might stimulate emotional engagement with, and a sense of connection to, persons distant in time and space. Writing about similar artefacts in the Alexander Girard Collection at the Museum of International Folk Art in Santa Fe, the anthropologist Doris Francis points out that, although they 'came from societies different to our own, all of them were accessible and immediately recognizable, intimate and familiar. They were about our shared human bodies, and we perceived them as our embodied selves.'[10] Being partial and anonymous, the votives stir our imaginations – we wonder who left them and whether their prayers were answered, filling the gaps according to our own experiences and assumptions. By recalling occasions when we too felt powerless and afraid for our own (or another's) health, we envisage the lives that touched them.

KEEPING THE FAITH

Another oddly compelling subset of objects in the Science Museum's Medicine collection is of protective amulets and charms. One of the storerooms is densely packed with countless trays of these typically small, unassuming objects that were carried or worn close to the skin (fig. 232). Almost every culture in the world has at some point used objects to ward off misfortune, or conversely to bring about good luck. Broadly speaking, amulets were intended to 'eliminate the negative' and charms to 'accentuate the positive'.[11] Formed of an astonishing variety of materials ranging from the bizarre to the

everyday, these objects have been transformed into the extraordinary by the beliefs and hopes invested in them by their former owners and users. Aside from their typically diminutive size, they are united only by their perceived magical or spiritual properties.

As well as constituting one of the Science Museum's most culturally diverse collections, these objects are perhaps also among the most affective. Part of the appeal is the sheer variety of their forms – a brooch made from an otter's paw, a crust of bread in a tiny muslin bag, a Heinz pickle pendant given away at the 1893 Chicago World's Fair, copper rings hewn from old pennies, and an embracing couple carved from Bolivian stone, to name but a few.

Some are better provenanced than others, having been collected originally by Edwardian folklorists including Edward Lovett (1852–1933), who kept detailed field notes relating to their place of origin and generic purpose. We know, for example, that the aforementioned penny rings were made and used by sailors to protect against rheumatism, dried mole's feet were carried to prevent cramp and toothache (fig. 233), and beaded glass necklaces were placed around the necks of infants in east London to ward off bronchitis (fig. 234). Lovett recorded little to no information, however, about the specific meanings ascribed to these objects by their former owners, claiming that few were prepared to discuss their personal beliefs: 'By talking of the devil things would go wrong'.[12]

Like the votives, the emotional power of these objects is magnified by their volume. Unlike the former, however, amulets and charms were never intended to be accumulated or displayed *en masse*. Instead, they constitute tangible expressions of private hopes and fears, and their uses and meanings are often highly subjective. In some – but by no means all – instances, their power is considered specific to an individual and cannot be transferred, such as the handwritten charms produced to order by an Ashanti shaman to provide personalised protection and insurance against illness (fig. 235).

Intended to be secreted about the person, amulets share an intimate relationship with the human body. In this way, they function almost as extensions of the self – objects relied upon to manage the uncertainty of daily life. Besides serving a practical purpose, their small scale

بسم الله الرحمن الرحيم ﴿ﻣﺤﺒﺔ﴾ ملك والناس كلهم والكبرين
والكبرة والرجل والمرأة وصغير وكبير والناس كلهم وجعاير
تكتب وتعلق وتنشر ﴿تجيب محبة الناس﴾ سريج عمر كبير

A666428

بسم الله الرحمن الرحيم اللهم صل على محمد وعلى آل
محمد وسلم اللهم صل على محمد وعلى آل محمد وسلم اللهم
صل على سيد نا محمد وعلى آل محمد وسلم اللهم صل على آل محمد
وسلم اللهم صل على محمد وعلى آل محمد وسلم
برس عليكم الشواظ من نار وغار سرو فلا فلا يرسل
عليكم شواظ من نار وغار سرو فلا فلا يرسل عليكم
شواظ من نار وغار سرو فلا فلا يرسل عليكم شواظ من
نار وغار سرو فلا فلا يرسل عليكم شواظ من نار وسرو
فلا فلا يرسل عليكم شواظ من نار وغار سرو فلا فلا
يرسل عليكم شواظ من نار وغار سرو فلا فلا يرسل
عليكم شواظ من نار وغار سرو فلا فلا يرسل عليكم
شواظ من نار وغار سرو فلا فلا يرسل عليكم شو
اظ من نار وغار سرو فلا فلا يرسل عليكم شواظ
من نار وغار سرو فلا فلا يرسل عليكم شواظ
من نار وغار سرو فلا فلا يرسل عليكم شواظ من
نار وغار سرو فلا فلا يا قيوم يا قيوم يا قيوم

might even aid their efficacy, given that, as Wellcome's most recent biographer Frances Larson puts it, 'miniature things can impart a feeling of human control and transcendence'.[13]

Occasionally their resonance is bound up with sentimental value, defined by the philosopher Guy Fletcher as something that has acquired its (emotional) significance via association with a significant person, place or memorable experience.[14] This could well be true of hundreds of amulets in the Science Museum's collection, but in the absence of accompanying documentation, we can only guess at the memories and relationships they may once have commemorated.

One unambiguous example is a beaded leather amulet in the form of a turtle, a creature associated with women's health and fertility in many Native American cultures (fig. 236). Gifted to female infants by the Sioux people of the Great Plains, they were initially used as cradle toys and later worn throughout childhood to protect against illness and premature death. Their most significant function, however, was to contain a small piece of the umbilical cord, representing the intergenerational bond between the newborn girl and her ancestors. Traditionally made by the child's mother or grandmother, they became family keepsakes and were usually kept throughout life, well beyond their 'useful' period of protection in early years. Baby boys received similar umbilical containers in the shape of lizards, which are associated with speed and rejuvenation. While long severed from its original context, this object and others like it nonetheless convey powerful, relatable emotions: the protective nature of maternal love; the desire to pass on to younger generations something special and unique; a parent's fierce hope that their child will not succumb to disease or accident.

Another of Lovett's collections – a group of amulets carried by soldiers fighting on the Western Front during the First World War (fig. 237) – is especially poignant. Unremarkable except by association, a small display of these objects was identified by visitors as one of the most memorable exhibits in a Science Museum exhibition entitled *Wounded: Conflict, Casualties and Care* (July 2016 – June 2018). The fact that we do not know who their individual owners were – or indeed whether or not they survived – hardly matters. The deep-seated human need to hold on, quite literally, to some semblance of hope in such bleak circumstances is immediately apparent.

Lovett himself had a complicated relationship with his life's work. Despite being outwardly sceptical and dismissive of their efficacy, he nonetheless produced, gave and sold amulets to others. The beneficiaries included his youngest son, for whom he made a charm to wear when he, too, was conscripted to fight in the First World War.[15]

These objects prompt us to consider our own methods of coping with life's unpredictability. Many people, for example, wear a special piece of jewellery gifted by a relative or friend, have a tattoo signifying a meaningful relationship, or carry some other small object, such as a photograph, to remind themselves of the loved ones and places that bring comfort and reassurance.

In the Science Museum's Medicine Galleries, mass displays of anonymous objects – including those discussed here – are complemented by six first-person stories expressed through a single or small group of objects, displayed individually and accompanied by recorded spoken testimony. In reuniting objects and narratives, these exhibits we hope provide an emotionally relatable 'way in' to the historic collection and the human experiences it represents.

One example that resonates particularly strongly with the amulets and charms is a hospital gown, created by the Liverpool-based artist Tabitha Moses (fig. 238). Entitled *Investment: Tabitha's Gown,* it is part of a series of three hand-embroidered garments, each of which carries a mandala-style composition to represent the individual journey of a woman undergoing in-vitro fertilisation (IVF) treatment – in this case the artist herself. All three gowns depict icons of intense personal significance, interspersed with the medical paraphernalia of infertility treatment.

Those facing infertility today have more options available to them than at any other time in history, yet IVF remains far more likely to fail than succeed, for reasons still not entirely understood. It is hardly surprising, then, that many fertility patients find themselves contemplating alternative therapies and even developing their own rituals in addition to receiving all the help that medical science can offer. Tabitha tried acupuncture, massage and hypnotherapy, and – despite not considering herself a religious person – even took to lighting a candle

Fig. 236
Turtle-shaped amulet,
Sioux people, 1880–1920.

Science Museum Group.
Object number A51675

Fig. 237
Amulets worn by men of
the London Regiment in the
First World War, from the
Lovett Collection; from left
to right: King Edward's hand,
fragment from an artillery
shell, 'lucky' black cat.

Science Museum Group.
Object numbers A79870,
A79904, A79871

Fig. 238
Tabitha Moses,
Investment: Tabitha's Gown,
embroidered thread and
mirrors on linen, 2014
(pictured with the maker).
Science Museum Group.
Object number 2016-568

Fig. 239
Milton Roy kidney machine
for home dialysis, *c*.1966.
Used by Moreen Lewis, one
of the first patients to have
a dialysis machine at home.
Science Museum Group.
Object number 1979-202

each evening in front of a Ghanaian fertility charm and a baby Jesus nativity figurine. Her gown highlights the fact that many of us employ a pluralistic approach when confronting the unpredictability of health matters, combining mainstream medicine with complementary treatments, elements of formal religion and folk beliefs. According to the 2002 National Health Interview Survey, as many as 36 per cent of adults in the United States use some form of alternative medicine, jumping to 62 per cent if you include prayer.[16] As Tabitha explains:

> Once the scientists have done their bit there follows 'the two week wait' to find out if the embryo has taken root. There is nothing to be done but hope and pray (if that's your cup of tea, and even if it isn't). Personal rituals and beliefs help us to live with the uncertainty. There is a need to feel in control of some part of the process.[17]

Tabitha's gown reflects her own highly personal experience of trying for a baby, right down to the prominent bloodstain, painstakingly reproduced

from a photograph of the bleeding that signalled the miscarriage of her first pregnancy. The labour of embroidering the gowns was in itself a coping mechanism: a means of guaranteeing at least one positive outcome. Yet the hope it embodies speaks to the anonymous objects in the wider gallery, encouraging visitors to wonder about and empathise with their former owners. Her voice will, in effect, stand in for theirs, enhancing these artefacts' imaginative potential.

TAKING A LEAP OF FAITH

The following two objects, like Tabitha's gown, carry an emotional charge that is enhanced – or even contingent upon – their individual narratives rather than their collective presence. At first glance, this rather drab-looking piece of medical equipment from the 1960s hardly seems like the most moving or affecting of artefacts (fig. 239). The size of a large chest of drawers, its veneered chipboard doors hinge open to reveal a complex system of dials and switches, under which sit a substantial tank and several

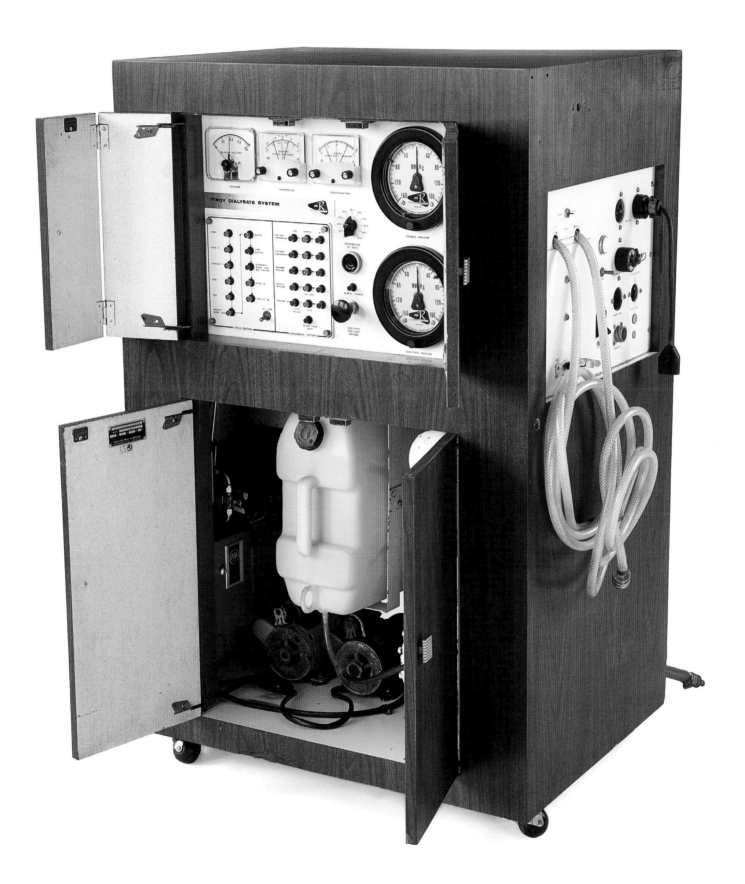

metres of plastic tubing. Only by reading the label do we learn that it is an early kidney-dialysis machine, designed to blend into the home environment.

In fact, this particular model was donated to the Science Museum by its user, Moreen Lewis, who in 1966 became one of the first British patients to have a dialysis machine at home. Nicknamed 'Dr Who', the machine kept her alive for nine years, replacing the function of her chronically diseased kidneys by filtering waste products from her blood. The procedure took ten hours and involved passing her entire blood volume through the machine three times a week. The knowledge of its story transforms this seemingly mundane object into an item of intensely personal significance.

In the 1970s it was rare for biographical details and histories of use to be recorded (unless the object in question belonged to an eminent scientist or public figure), but on this occasion, we were lucky to receive a detailed account of Moreen's story along with her machine, narrated by the journalist Keith Bill in his 1968 book *Plug in For Life*. While dialysis presented the

only hope of survival for patients like Moreen, the prohibitive costs of the equipment and associated care meant that only a fraction could be treated on the relatively new National Health Service. As a single 39-year-old without dependants, she was not considered eligible for treatment and, drawing on her renewed Christian faith, eventually accepted her terminal diagnosis. The turning point came when a group of friends rallied together to secure Moreen a place at the private National Kidney Centre in Finchley, established in 1966 by Stanley Shaldon (1931–2013) – a nephrologist and pioneering proponent of home dialysis. Shaldon believed that, if patients could be trained to operate their machine autonomously at home, they would benefit from a greater degree of independence, while simultaneously freeing up staff and beds in renal units (fig. 240). As his hospital received no state funding, patients had to pay £7,000 upfront for their own machine and care – the average price of two houses in 1965. Moreen was fortunate enough to have a wealthy uncle who covered her costs. Other patients resorted to fundraising campaigns run via local newspapers or church

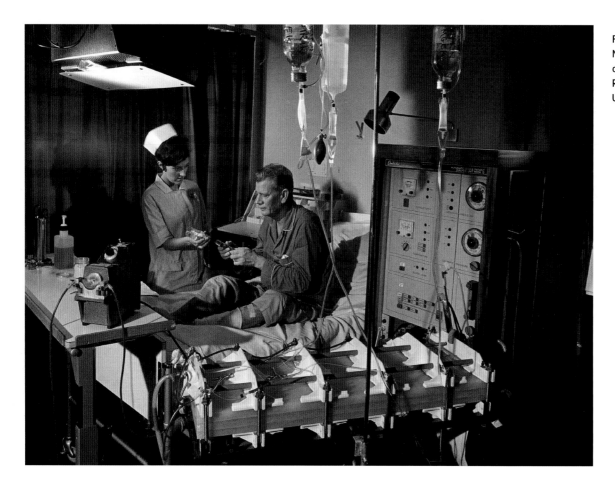

Fig. 240
Nurse tending to a patient on a dialysis machine, Royal Free Hospital Renal Unit, London, 1968.

groups (akin to the more recent internet crowd-funding phenomenon).

Paying for the machine was just half the battle, however. Allowing patients and their families to manage these complex blood-filtering machines unsupervised was controversial. Freedom brought with it an enormous amount of responsibility – not only for setting up the machine correctly but also handling potentially life-threatening leaks and breakdowns outside a medical setting. The burden proved too much for some, resulting in relationship breakdowns and even suicide. While Moreen ultimately adapted well to home dialysis, her mother and primary carer, Charlotte, found the months of training arduous and the pair encountered their share of faults, along with water and blood leaks. Charlotte admitted: 'When I used to watch Moreen on the machine during those early months… I sometimes wondered whether it was all worthwhile. Had we really done the right thing?'[18]

As earlier chapters have shown, anonymous medical relics are certainly capable of producing and transmitting emotion (the iron lungs and prosthetic limbs being some of the most affecting objects in the Science Museum's collection). Presenting this object as Moreen's 'Dr Who', however, sets up an immediate empathic connection to a machine with which she shared a personal relationship. Hearing Moreen's story, we come away with a greater appreciation of the leap of faith required of both patients and their clinicians to make home dialysis a reality. Almost all medical interventions involve a degree of risk, and outcomes are rarely certain. When deciding on a course of treatment, therefore, everyone involved must trust that the physical, emotional and financial demands are worth it for the possibility of a longer or better quality of life.

The careful balancing of risk versus benefit in medical decision-making is similarly embodied in a bespoke Perspex mask, produced for a patient undergoing radiotherapy for head and neck cancer at St Luke's Hospital in Guildford in the mid-1960s (fig. 241). Masks like these, known also as moulds, are used to keep the patient's body fixed in exactly the same position throughout their six-week course of treatment. They allow radiotherapy beams to be targeted directly at the cancerous cells, while minimising damage to the surrounding healthy tissue. Many people – quite understandably – find the

experience of wearing the mask claustrophobic, which may be compounded by feelings of anxiety, depression and fatigue that are common during radiotherapy treatment. It almost provokes a visceral reaction – it is difficult not to imagine the sensation of the hard plastic pressed against your own cheek.

While we cannot say who this particular mask belonged to, both it and the cast from which it was created are literally marked by individual human experience. In their disembodied state, they act as ghostly relics of a difficult period in that person's life, the outcome of which we will never know. Despite not being catalogued as 'human remains', it undoubtedly serves as a singular, irreducible link to the patient whose face it immortalises.

Since they are tailor-made to each individual, the radiotherapy moulds worn during treatment are routinely offered to patients once it is completed. While relatively few accept, some do opt to keep them – whether as a badge of pride, memento of everything they have gone through, reminder of the fragility of life or merely to take satisfaction in destroying them. Knowing this, the Science Museum approached Guy's Hospital in London, seeking radiotherapy patients who would be willing to donate their mask and story for inclusion in the Medicine Galleries. Recognising the latent poignancy of these objects, we wished to collect and present supporting personal testimony alongside the physical artefacts, revealing what they actually *felt* like to wear, how the individuals concerned managed their discomfort (and the experience of being a cancer patient), and how they and their doctors reached decisions about their programme of care. In so doing, we hoped – as with Moreen's kidney machine and Tabitha's gown – to establish an empathic connection between visitor, object and human subject.

All three items are indicative of a wider cultural shift in the Museum's approach to collecting, away from 'idealist, internalist accounts of sequences' and towards 'more nuanced approaches [which] have come to look at science as part of social life'.[19] Since the late 1980s, far less importance has been ascribed to an object's meaning within a series and far more to its meaning within a story. What and whose story is being told will of course vary, but a major consequence for the collection at the Science Museum has

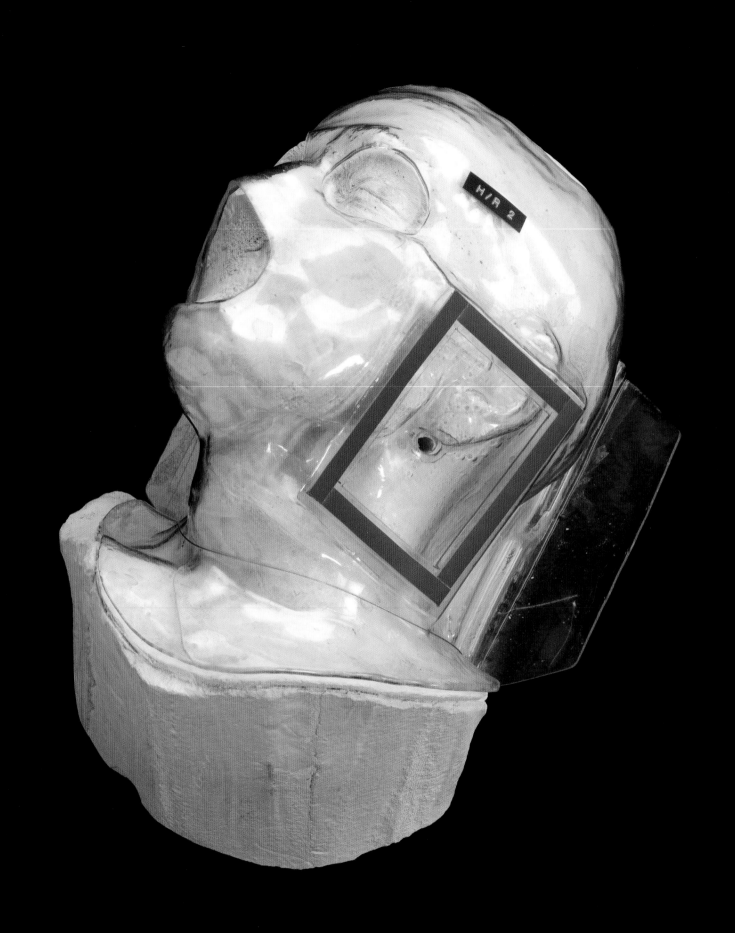

Fig. 241
Face mould for
administrating radiotherapy
and original plaster cast,
St Luke's Hospital,
Guildford, c. 1965.

Science Museum Group.
Object number 1990-112/6

been a greater number of acquisitions that reveal the patient's experience of medicine.

Our knowledge and understanding of the human body – and what to do when things go wrong – has been transformed in recent years, yet faith and feeling remain central to our encounters with ill health. Technological advances in medicine have brought with them fresh ambiguities and ethical dilemmas, further complicating decision-making around benefit versus risk and determining when to withdraw or withhold treatment. As the surgeon and writer Atul Gawande has astutely observed:

> The core predicament of medicine – the thing that makes being a patient so wrenching, being a doctor so difficult, and being part of a society that pays the bills so vexing – is uncertainty. With all that we know nowadays about people and diseases and how to diagnose and treat them, it can be hard to see this, hard to grasp how deeply uncertainty runs.[20]

Far from being relegated to the past, beliefs play a fundamental role in helping people make sense of illness and cope with its – and modern medicine's – inherent unpredictability. Moreover, it is rarely a case of either/or when it comes to investing one's trust in a being or philosophy; science, people and religion can all provide peace of mind, and are by no means mutually exclusive.

The objects and stories discussed here expose this messiness, revealing the lived experience of sickness and health. Emotional objects facilitate the transmission of empathy, helping to bridge the temporal, cultural and geographical distances between their creators and users and today's gallery visitors. Their value relates less to their rarity and physical condition and more to their ability to illuminate what it *feels* like to encounter a threat to the health of our minds or bodies.

We have seen how objects separated by thousands of years are capable of absorbing and transmitting emotion, albeit in very different ways. In the case of the votives and amulets, knowledge of their purpose and function transforms them from vestigial relic to imaginative portal – a (murky) window into other lives. Their power to elicit feeling is enhanced by their collective presence. For the stand-alone objects, it is narrative that elevates them from the mundane, establishing a human connection that prompts us to consider how faith, hope and fear might operate in relation to our own experiences of illness, treatment and recovery.

NOTES

1

COLLECTING MEDICINE
ROUTES AND ROOTS OF MEDICINE AT
THE SCIENCE MUSEUM

1. See S. Alberti, 'Objects and the Museum', *Isis*, vol. 96, pp. 559–71 and Simon Schaffer, 'A Science whose Business is Bursting' in L. Dalston (ed.), *Things that Talk: Object Lessons from Art and Science* (Zone Books, 2007) for opposing arguments on whether objects can talk.
2. Burroughs Wellcome & Co. became the Wellcome Foundation Ltd in 1924. In 1986 the Wellcome Foundation Ltd became Wellcome plc, and nine years later merged with Glaxo to form Glaxo Wellcome. In 2000 GlaxoWellcome merged with SmithKline Beecham to form GlaxoSmithKline.
3. R. Church and E. M. Tansey, *Burroughs Wellcome & Co.: Knowledge, Trust, Profit and the Transformation of the British Pharmaceutical Industry, 1880–1940* (Crucible Books, 2007), p. 44.
4. F. Larson, *An Infinity of Things: How Sir Henry Wellcome Collected the World* (Oxford University Press, 2009), p. 10.
5. Burroughs Wellcome & Co., *The Romance of Exploration and Emergency First-Aid from Stanley to Byrd: Chicago Century of Progress Exposition* (Burroughs Wellcome & Co., 1934).
6. Larson, *An Infinity of Things*, p. 9.
7. Ibid.
8. Ibid., p. 10.
9. Larson, *An Infinity of Things*; see Larson's chapter 8 for information on agents working in India.
10. Wellcome Library Archives: WHMM/RP/Jst/B/1.
11. Letter from Sir John Samuel to Peter Johnston-Saint, 11 June 1926. Wellcome Library Archives: WAHMM/RP/Jst/B/1.
12. H. Turner, *Wellcome: The Man, the Collection, the Legacy* (Heinemann Educational Publishers, 2001). p. 40.
13. Johnston-Saint reports, January–March 1932, Wellcome Library Archives: WA/HMM/RP/Jst/B.10.
14. Ibid.
15. Johnston-Saint reports, January–March 1932, Wellcome Library Archives: WA/HMM/RP/Jst/B.10.
16. Wellcome Library Archives: WA/HMM/St/lat/A.115. Letter to Mrs Johnston-Saint, 16 June 1924.
17. Wellcome Library Archives: WA/HMM/CM/Col/44.
18. Private Purchase Book, Wellcome Library Archives: WA/HMM/FI/Pri/4.
19. Reminiscences of staff, Wellcome Library Archives: WF/M/H/07/03.
20. Larson, *An Infinity of Things*, pp. 213–17.
21. Borer's resignation letter, Wellcome Library Archives: WA/HMM/St/Lat/A.26.
22. G. Russell, 'The Wellcome Historical Medical Museum's Dispersal of Non-Medical Material 1936–1983', *Museums Journal Supplement*, vol. 86, supplement 1986, details the transfer and sale of items.
23. Reminiscences of staff, Wellcome Library Archives: WF/M/H/07/03.
24. SMG Collection Development Strategy, 2016. The Science Museum Group consists of the Science Museum in London, the National Railway Museum in York, National Museum of Science and Media in Bradford, Museum of Science and Industry in Manchester and Locomotion: the National Railway Museum in Shildon.
25. See T. Stephens and R. Brynner, *Dark Remedy: The Impact of Thalidomide and its Revival as a Vital Medicine* (Basic Books, 2001) on the background and development of thalidomide.
26. Oral history, recorded by the Thalidomide Society with Tracey Baynam, October 2016.
27. Ibid.
28. Stewart Emmens in conversation with the author, May 2018.

2

PROSTHESES AT THE SCIENCE MUSEUM
PEG-LEGS, PYLONS AND THE PIANIST'S ARM

1. Letter from Stevenson to Henley, May 1883 in Sidney Colvin (ed.), *The Letters of Robert Louis Stevenson*, vol. II (1880–87) (Methuen & Co., 1911).
2. Information recorded on the original catalogue entry of this object by Dr Ian Fletcher, Senior Medical Officer at Queen Mary's Hospital, Roehampton.
3. G. E. Marks, *A Treatise on Marks' Patent Artificial Limbs with Rubber Hands and Feet* (New York, 1888), p. 111.
4. Aimee Mullins, https://www.ted.com/talks/aimee_mullins_prosthetic_aesthetics.

3

THE PROBLEMATIC BODY
DISSECTION AND THE RISE OF THE ANATOMY SCHOOL

1. G. Ferrari, 'Public Anatomy Lessons and the Carnival: The Anatomy Theatre of Bologna', *Past and Present*, no. 117 (1987), pp. 50–106.
2. S. Chaplin, 'Dissection and Display in Eighteenth-Century London', in Piers Mitchell (ed.), *Anatomical Dissection in Enlightenment England and Beyond: Autopsy, Pathology and Display* (Ashgate, 2012), p. 97.
3. F. J. Knox, *The Anatomist's Instructor, and Museum Companion; being Practical Directions for the Formation and Subsequent Management of Anatomical Museums* (Adam and Charles Black, 1836), p. 3.
4. Quoted in Chaplin, 'Dissection and Display in Eighteenth-Century London', p. 101.
5. Quoted in R. Richardson, *Death, Dissection and the Destitute* (Phoenix Press, 2001, 2nd edn), p. 197.
6. Ibid., p. xv.
7. Quoted in ibid., p. 49.
8. Quoted in R. Mandressi, 'Affected Doctors: Dead Bodies and Affective and Professional Cultures in Early Modern European Anatomy', in *Osiris*, vol. 31, no. 1 (2016), p. 121.
9. Quoted in ibid., p. 123.
10. Quoted in Richardson, *Death, Dissection and the Destitute*, pp. 30–31.
11. J. H. Warner and J. M. Edmonson, *Dissection: Photographs of a Rite of Passage in American Medicine, 1880–1930* (Blast Books, 2009), pp. 8–10.
12. Human Tissue Act, 2004.

4

THE CHILD IN THE IRON LUNG
POLIO, PUBLIC HEALTH AND PROTECTION

1. E. Sass, *Polio's Legacy: An Oral History* (University Press of America, 1996), p. xiii.
2. T. Gould, *A Summer Plague: Polio and its Survivors* (Yale University Press, 1995), p. 232.
3. Sass, *Polio's Legacy*, p. 79.
4. Ibid., p. 74.
5. Ibid., p. 117.
6. Gould, *A Summer Plague*, p. 91.
7. T. M. Daniel and F. Robbins, *Polio* (University of Rochester Press, 1997), p. 55.
8. Ibid., p. 32.
9. See *Marshall Evening Chronicle*, 9 January 1956.
10. Ibid.
11. Sass, *Polio's Legacy*, p. 55.
12. Ibid.
13. Joni Mitchell, interview with Pamela Wallin, *Pamela Wallin Live*, CBC TV, 19 February 1996.
14. Quoted in Daniel and Robbins, *Polio*, p. 104.
15. Quoted in G. McKay, '"Crippled with Nerves": Popular Music and Polio, with Particular Reference to Ian Dury', 2009, University of Salford, http://usir.salford.ac.uk/2372.
16. Available online.
17. Ian Dury and the Blockheads, 'Hey, Hey, Take Me Away', from *Laughter* (Stiff Records), 1980.
18. G. Williams, *Paralysed with Fear: The Story of Polio* (Palgrave, 2015), passim.
19. *Hospital School*, 1945 (dir. Arthur Barnes), film.britishcouncil.org.

20. *His Fighting Chance*, 1946, UK Central Office of Information.
21. Footage available from Special Collections in Mass Media and Culture, University of Maryland Libraries, Washington DC.
22. J. Davies, 'Meet Debra, the Historic Collecting Box', Science Museum blog, 1 March 2018.
23. Rebecca Hardy, *Mail Online*, 9 August 2013.
24. Roisin Tierney, in conversation with the author, 24 May 2018.
25. Ibid.
26. Mrs Blackstone, letter to the Science Museum, object number 1989-832.

5
CREATED THROUGH CONFLICT
THE DEVELOPMENT OF MILITARY MEDICINE

1. Quoted in Dr J. Wright, *A History of War Surgery* (Amberley, 2011), p. 14.
2. E. M. Wrench, 'The Lessons of the Crimean War', *British Medical Journal*, vol. II, no. 2012 (July 1899), p. 206.
3. Ibid., p. 205.
4. 'The Sick and Wounded Fund', *The Times* (8 February 1855), p. 7.
5. 'Miss Nightingale, in the Hospital, at Scutari', *Illustrated London News* (24 February 1855), p. 176.
6. J. E. McCallum, *Military Medicine from Ancient Times to the 21st Century* (ABC-CLIO, 2008), p. 50.
7. J. Symons, *Buller's Campaign* (House of Stratus, 2008), p. 132.
8. 'The Royal Commission on South African Hospitals', *British Medical Journal*, vol. 1, no. 2091 (1901), p. 240.
9. M. Borden, *The Forbidden Zone: A Nurse's Impression of the First World War* (Hesperus Press, 2008), pp. 95–96.
10. L. Byrski, 'Emotional Labour as War Work: Women up Close and Personal with McIndoe's Guinea Pigs' in *Women's History Review*, vol. 21, issue 3 (July 2012), p. 342.

6
PREVENTING CERVICAL CANCER
VAGINAS, VACCINATION AND VENEREAL DISEASE

1. 'Cancer Statistics for the UK', Cancer Research UK, https://www.cancerresearchuk.org/health-professional/cancer-statistics-for-the-uk, accessed 13 August 2018.
2. J. Lorber, *Gender and the Social Construction of Illness* (AltaMira Press, 2000).
3. O. Moscucci, *The Science of Woman: Gynaecology and Gender in England, 1800–1929* (Cambridge University Press, 1990).

4. J. Harsin, *Policing Prostitution in Nineteenth-Century Paris* (Princeton University Press, 1985).
5. 'The Prostitute whose Pox Inspired Feminists', 20 July 2017, Wellcome Collection, https://wellcomecollection.org/articles/outsiders-the-prostitute, accessed 13 August 2018.
6. L. Hall, 'Venereal Diseases and Society in Britain, from the Contagious Diseases Acts to the National Health Service', in R. Davidson and L. A. Hall (eds), *Sex, Sin and Suffering: Venereal Disease and European Society Since 1870* (Routledge, 2001).
7. M. Spongberg, *Feminizing Venereal Disease: The Body of the Prostitute in Nineteenth-Century Medical Discourse* (Macmillan Press, 1997).
8. J. C. A. Recamier, *Recherches sur le traitement du cancer* (Gabon, 1829).
9. Quoted in I. Lowy, *A Woman's Disease: The History of Cervical Cancer* (Oxford University Press, 2011), p. 129.
10. Ibid., p. 44.
11. D. E. Carmichael, *The Pap Smear: Life of George N. Papanicolaou* (Charles C. Thomas Publisher, 1973).
12. M. J. Casper and A. E. Clarke, 'Making the Pap Smear into the "Right Tool" for the Job: Cervical Cancer Screening in the USA, circa 1940–95', *Social Studies of Science*, vol. 28, no. 2 (April 1998), pp. 259–61.
13. Dr G. N. Papanicolaou, 'New Cancer Diagnosis', presented at National Conference of Race Betterment; Race Betterment Foundation, Michigan, 2–6 January 1928 (Race Betterment Foundation, 1928), pp. 528–34.
14. G. N. Papanicolaou and H. F. Traut, 'The Diagnostic Value of Vaginal Smears in Carcinoma of the Uterus,' *American Journal of Obstetrics and Gynecology*, vol. 42, no. 2 (August 1941), pp. 193–206.
15. G. A. Vilos, 'The History of the Papanicolaou Smear and the Odyssey of George and Andromache Papanicolaou', *Obstetrics & Gynecology*, vol. 91, no. 3 (March 1998), pp. 479–83.
16. H. Speert, 'Papanicolau, George Nicholas', in Charles Coulston Gillispie (ed.), *Complete Dictionary of Scientific Biography* (Charles Scribner's Sons, 2008), pp. 291–2.
17. L. Bryder, 'Debates about Cervical Screening: An Historical Overview,' *Journal of Epidemiology and Community Health*, vol. 62, no. 4 (May 2008), pp. 284–7.
18. H. C. McLaren, *The Prevention of Cervical Cancer* (Charles C. Thomas, 1964), p. 96.
19. 'The NHS Constitution for England', Department of Health and Social Care, https://www.gov.uk/government/publications/the-nhs-constitution-for-england/the-nhs-constitution-for-england, accessed 13 August 2018.

20. NCPCC newsletter (January 1966), Wellcome Library Archives: SA/SMO.L.35.
21. L. A. Reynolds and E. M. Tansey (eds), *History of Cervical Cancer and the Role of the Human Papillomavirus, 1960–2000: Wellcome Witness to Twentieth-Century Medicine*, vol. 38 (Wellcome Trust, 2009), p. 22.
22. Casper and Clarke, 'Making the Pap Smear into the "Right Tool" for the Job', p. 269.
23. Reynolds and Tansey (eds), *History of Cervical Cancer and the Role of the Human Papillomavirus*, p. 10.
24. Ibid., pp. 11–12.
25. Ibid., pp. 12–13.
26. E. Lynch-Farmery, 'Cervical Cancer Screening in the United Kingdom, 1986 to 1996: A Decade of Change', *Journal of Obstetrics and Gynaecology Canada*, vol. 18, no. 2 (December 1996), pp. 1251–9.
27. 'Guidance on the Use of Liquid-Based Cytology for Cervical Screening: Technology Appraisal Guidance', National Institute for Health and Care Excellence, nice.org.uk/guidance/ta69, last modified 22 October 2003.
28. J. Scotto and J. C. Bailar, 'Rigoni-Stern and Medical Statistics: A Nineteenth-Century Approach to Cancer Research,' *Journal of the History of Medicine and Allied Sciences*, no. 1 (January 1969), pp. 65–75.
29. V. Beral, 'Cancer of the Cervix: A Sexually Transmitted Infection?' *Lancet*, vol. 25, no. 1 (May 1974), pp. 1037–40.
30. J. L. Grimes, 'HPV Vaccine Development: A Case Study of Prevention and Politics', *Biochemistry and Molecular Biology Education*, vol. 34, no. 2 (March 2006), pp. 148–54; M. H. Schiffman, P. Castle, 'Epidemiologic Studies of a Necessary Causal Risk Factor: Human Papillomavirus Infection and Cervical Neoplasia', *Journal of the National Cancer Institute*, vol. 95, no. 6 (March 2003), E2; F. Bosch, M. Manos, N. Munoz, M. Sherman, A. Jansen, J. Peto, M. Schiffman, V. Moreno, R. Kurman, K. Shah, 'Prevalence of Human Papillomavirus in Cervical Cancer: A Worldwide Perspective, International Biological Study on Cervical Cancer (IBSCC) Study Group', *Journal of the National Cancer Institute*, vol. 87, no. 11 (June 1995), pp. 796–802.
31. 'Human Papillomavirus (HPV) and Cervical Cancer', World Health Organisation, http://www.who.int/news-room/fact-sheets/detail/human-papillomavirus-(hpv)-and-cervical-cancer, accessed 13 August 2018.
32. As with any vaccination, a number of competing HPV vaccines were developed by major pharmaceutical companies. The government is responsible for choosing which vaccines are purchased for national immunisation, based on an assessment of cost, risks and benefits.

GlaxoSmithKline's Ceravix vaccine, which protects against two types of HPV (16 and 18), was selected for use in the UK from September 2008 to August 2012. Following 'a competitive tendering exercise', in September 2012 the vaccine was changed to Merck's Gardasil vaccine, which protects against four types of HPV (6, 11, 16 and 18). The capitalist implications of vaccination decision-making are beyond the scope of this chapter, but it is an important point that has been made elsewhere.

33. 'HPV Vaccine', National Health Service, https://www.nhs.uk/conditions/vaccinations/ hpv-human-papillomavirus-vaccine/, accessed 13 August 2018.

34. 'Head and Neck Cancers Statistics', Cancer Research UK, https://www.cancerresearchuk. org/health-professional/cancer-statistics/ statistics-by-cancer-type/head-and-neck-cancers, accessed 13 August 2018.

35. This age limit was selected because most people get HPV when they first become sexually active. Evidence suggests that the vast majority of people who contract HPV get the infection before the age of 45.

36. 'Joint Committee on Vaccination and Immunisation Statement on HPV Vaccination of Men who Have Sex with Men', Public Health England, November 2015.

37. 'HPV Vaccination Programme for Men who Have Sex with Men (MSM): Clinical and Operational Guidance,' Public Health England, April 2018.

38. 'Statement on HPV Vaccination: Joint Committee on Vaccination and Immunisation,' Public Health England, July 2018.

39. 'Cancer Statistics for the UK,' Cancer Research UK, http://www.cancerresearchuk.org/health-professional/cancer-statistics-for-the-uk, accessed 13 August 2018.

40. 'Guidance: Health Matters: Making Cervical Cancer Screening More Accessible,' Public Health England, https://www.gov.uk/ government/publications/health-matters-making-cervical-screening-more-accessible/ health-matters-making-cervical-screening-more-accessible-2, accessed 13 August 2018; 'Cervical Screening in the Spotlight: An Audit of Activities Undertaken by Local Authorities and Clinical Commissioning Groups to Increase Cervical Screening Coverage in England,' Jo's Cervical Cancer Trust, https://www.jostrust.org.uk/sites/default/files/ cervical_screening_in_the_spotlight_-_final.pdf, accessed 13 August 2018.

41. 'Barriers to Cervical Screening among 25–29 Year Olds', Jo's Cervical Cancer Trust, https://www.jostrust.org.uk/sites/default/ files/ccpw17_survey_summary.pdf, accessed 13 August 2018.

42. M. Lohan and W. Faulkner, 'Masculinities and Technologies: Some Introductory Remarks', *Men and Masculinities,* vol. 6, no. 4 (April 2004), pp. 319–29.

7

HEARING DISTANT VOICES
PSYCHIATRY, ASYLUMS AND THE LIMITS OF HISTORY

1. Cassel Hospital, 'Report Review of Modern Treatment: Electric Shock Therapy' (Cassel Hospital, 1944), p. 3.

2. S. A. Paterson, 'Electric Convulsion Therapy', *British Medical Journal,* vol. 3, no. 5614 (August 1968), p. 375.

3. Ibid., p. 374.

4. W. Sargant and E. Slater, *An Introduction to Physical Methods of Treatment in Psychiatry* (Livingston, 1944), p. 141.

5. H. Bourne, 'The Insulin Myth', vol. 262, no. 6793 (November 1953), p. 964.

6. T. P. Rees, 'Back to Moral Treatment and Community Care', *Journal of Mental Science,* vol. 103, no. 431 (April 1957), p. 313.

7. H. Spandler, *Asylum to Action: Paddington Day Hospital, Therapeutic Communities and Beyond* (Jessica Kingsley, 2006), pp. 52–62.

8. D. Trefgarne, 'Care in the Community', House of Lords Debate (written answers), vol. 440 (14 March 1983), cc591-2wa.

8

THE PHARMACY SHOP
'A GREAT RAGE FOR MAHOGANY, VARNISH AND EXPENSIVE FLOOR-CLOTH'

1. J. Bell, 'Memoir of John Bell', *Pharmaceutical Journal and Transactions,* vol. 8 (1848–9), p. 591.

2. Slogan from an ongoing campaign by the Royal Pharmaceutical Society, 2000s.

3. P. Wallis, 'Apothecaries and Medicines in Early Modern London', in L. Hill Curth (ed.), *From Physick to Pharmacology: Five Hundred Years of British Drug Retailing* (Ashgate, 2006), p. 23.

4. A. Withey, *Physick and the Family: Health, Medicine and Care in Wales, 1600–1750* (Manchester University Press, 2011), p. 109.

5. M. Rowe and G. E. Trease, 'Thomas Baskerville, Elizabethan Apothecary of Exeter', in *Transactions of the British Society for the History of Pharmacy,* vol. 1, no. 1 (1970), pp. 16–20.

6. Quoted in L. Collingham, *The Taste of Empire* (Basic Books, 2017), p. 79.

7. J. Quincy, *A Compleat English Dispensatory: How Britain's Quest for Food Shaped the Modern World* (A. Bell, 1718), p. 166.

8. Withey, *Physick and the Family,* p. 114.

9. J. Burnby, 'House and Home for Apothecaries and Druggists,' *Pharmaceutical Historian,* vol. 8, no. 3 (1978), p. 2.

10. P. Wallis, 'Apothecaries and Medicines in Early Modern London', p. 23ff; P. Wallis, 'Consumption, Retailing, and Medicine in Early-Modern London,' *Economic History Review,* vol. 61, no. 1 (2007), passim.

11. J. Burnby, 'English Apothecaries and Probate Inventories', *Pharmaceutical Historian,* vol. 27, no. 4 (1997), p. 58.

12. Wallis, 'Consumption, Retailing and Medicine in Early-Modern London', p. 35.

13. Ibid., pp. 32–3, 38.

14. Ibid., p. 45.

15. W. Shakespeare, *Romeo and Juliet,* act v, scene 1, lines 42–4.

16. C. Merrett, *A Short View of the Frauds, and Abuses Committed by Apothecaries* (1670), p. 33.

17. S. W. F. Holloway, *Royal Pharmaceutical Society of Great Britain (1841–1991): A Political and Social History* (Pharmaceutical Press, 1991), p. 39.

18. J. K. Crellin and J. R. Scott, *Glass and British Pharmacy 1600–1900: A Survey and Guide to the Wellcome Collection of British Glass* (Wellcome Institute of the History of Medicine, 1972), p. 9.

19. Quoted in B. Hudson, *The School of Pharmacy, University of London: Medicines, Science and Society 1842–2012* (Academic Press, 2013), p. 3.

20. Quoted in H. Marland, '"The Doctor's Shop": The Rise of the Chemist and Druggist in Nineteenth-century Manufacturing Districts', in L. Hill Curth, *From Physick to Pharmacology: Five Hundred Years of British Drug Retailing* (Ashgate, 2006), p. 80.

21. *Pharmaceutical Journal and Transactions,* no. 3 (1843–4), p. 101.

22. Holloway, *Royal Pharmaceutical Society of Great Britain (1841–1991),* pp. 73–7.

23. Quoted in Crellin and Scott, *Glass and British Pharmacy 1600–1900,* p. 3.

24. Ibid., p. 12.

25. Quoted in J. Burnby, 'Pharmacy in the Mid-Nineteenth Century', *Pharmaceutical Historian,* vol. 22, no. 2 (1992), p. 4.

26. 'Women in British Pharmacy', *Chemist and Druggist* (July 1916), p. 792.

27. Quoted in P. Homan, B. Hudson and R. Rowe, *Popular Medicines* (Pharmaceutical Press, 2008), pp. 95–6.

28. 'Women in British Pharmacy', p. 792.

29. Quoted in Crellin and Scott, *Glass and British Pharmacy 1600–1900,* p. 9.

30. J. K. Crellin, 'Pharmacies as General Stores in the 19th Century', *Pharmaceutical Historian,* vol. 9, no. 1 (1979), pp. 5–6.

31. 'Pharmacalia', *Chemist and Druggist* (January 1882), p. 2.

32. P. M. Worling, 'Pharmaceutical Wholesale Distribution in the United Kingdom, 1950–1990', *Pharmaceutical Historian*, vol. 28, no. 4 (1998), p. 53.

33. S. C. Anderson, 'Community Pharmacy and Public Health in Great Britain 1936 to 2006: How a Phoenix Rose from the Ashes', *Journal of Epidemiology and Community Health*, vol. 61 (2007), p. 845.

34. S. C. Anderson, 'From "Bespoke" to "Off-the-Peg": Community Pharmacists and the Retailing of Medicines in Great Britain 1900 to 1970', *Pharmacy in History*, vol. 50, no. 2 (2008), pp. 58–9.

35. Anderson, 'Community Pharmacy and Public Health in Great Britain 1936 to 2006', p. 845.

36. Ibid.

9
MODELLING LIFE
EXPLORING THE HUMAN BODY

1. In the context of the four humours theory, a sanguine person is defined as lively, energetic and robust. A choleric person would be quick to anger, irritable and jealous. A melancholic is characterised by being passive, withdrawn and melancholic. Finally, a phlegmatic person would be passive and emotionless.

2. L. Jackson, 'From Atoms to Patterns: The Story of the Festival Pattern Group', *From Atoms to Patterns: Crystal Structure Designs from the 1951 Festival of Britain* (Wellcome Collection, 2008), p. 7.

3. Interview with James Watson, 2018, Science Museum blog, https://blog.sciencemuseum.org.uk/why-the-double-helix-is-still-relevant/

4. J. Ebenstein, 'Henry Wellcome's Anatomical Venus', 25 November 2014, Wellcome Collection blog, https://wellcomecollection.wordpress.com/2014/11/25/henry-wellcomes-anatomical-venus.

5. J. Ebenstein, *The Anatomical Venus* (Thames & Hudson, 2016), p. 24.

6. K. D. Hussey, 'Seen and Unseen: The Representation of Visible and Hidden Disease in the Waxworks of Joseph Towne at the Gordon Museum,' in *Interdisciplinary Studies in the Long Nineteenth Century*, vol. 19, no. 24 (2017), n.p. DOI: http//doi.org/10.16995/ntn.787

7. G. Ferry, *Dorothy Hodgkin: A Life* (Granta Books, 1998), p. 191.

8. Ibid., p. 213.

9. A. Cunningham, 'Preservation: Preserving Bodies Better and Longer', in *The Anatomist Anatomis'd: An Experimental Discipline in Enlightenment Europe* (Ashgate, 2010), p. 236.

10. S. de Chadarevian and N. Hopwood, 'Models and the Making of Molecular Biology', in *Models: The Third Dimension of Science* (Stanford University Press, 2004), p. 341.

11. S. Jeffries, 'The Naked and the Dead', *Guardian*, 19 March 2002.

12. https://bodyworlds.com/

13. http://marcquinn.com/artworks/body-alteration

14. Jackson, 'From Atoms to Patterns: The Story of the Festival Pattern Group', p. 26.

10
EMOTIONAL OBJECTS
FAITH AND FEELING IN THE MEDICINE COLLECTION

1. Quoted in *Bristol Medico-Chirurgical Journal*, vol. 22, no. 83 (1904), p.82.

2. Cited in G. M. Skinner, 'Sir Henry Wellcome's Museum for the Science of History', *Medical History*, vol. 30, no. 4 (1986), p. 403.

3. D. E. Moerman, *Meaning, Medicine and the 'Placebo Effect'* (Cambridge University Press, 2002).

4. R. Richardson, 'Human Remains', in Ken Arnold and Danielle Olson (eds), *Medicine Man: The Forgotten Museum of Henry Wellcome* (British Museum Press, 2003), p. 323.

5. J. Hughes, 'Fragmentation as Metaphor in the Classical Healing Sanctuary', *Social History of Medicine*, vol. 21, no. 2 (2008), pp. 217–36.

6. R. Jackson, *Doctors and Diseases in the Roman Empire* (British Museum Press, 1988), p. 157.

7. Ibid., p. 160.

8. L. Ricciardi, 'Canino (Viterbo): Il santuario etrusco di Fontanile di Legnisina a Vulci: Relazione delle campagne di scavo 1985 e 1986: L'altare monumentale e il deposito votivo', in *Notizie degli scavi di Antichità*, vol. 42, no. 3 (1988), pp. 137–209.

9. These small hand-painted plaques, named after the Latin for 'from a vow', gained prominence in Europe and beyond from the 16th century. Left in chapels in gratitude for answered prayers, they recount stories of cure or delivery from disaster through the intervention of a Christian patron saint.

10. D. Francis (ed.), *Faith and Transformation: Votive Offerings and Amulets from the Alexander Girard Collection* (Museum of New Mexico Press, 2007), p. 9.

11. T. Cadbury, 'The Charms of Scarborough, London, Etc.: The Collecting Networks of Charles Clarke and Edward Lovett', *Journal of Museum Ethnography*, vol. 25 (2012), p. 119.

12. E. Lovett, cited in J. M. Hill, *Cultures and Networks of Collecting: Henry Wellcome's Collection* (unpublished PhD thesis, Royal Holloway, University of London, 2004), p. 209.

13. F. Larson, 'The Things about Henry Wellcome', *Journal of Material Culture*, vol. 15, no. 1 (2010), p. 93.

14. G. Fletcher, 'Sentimental Value', *Journal of Value Inquiry*, vol. 43, no. 1 (2009), p. 56.

15. J. Hill, 'The Story of the Amulet: Locating the Enchantment of Collections', *Journal of Material Culture*, vol. 12, no. 1 (2007), p. 82.

16. P. Barnes, E. Powell-Griner, K. McFann, R. Nahin. CDC Advance Data Report no. 343, *Complementary and Alternative Medicine Use Among Adults: United States, 2002* (May 2004).

17. T. Moses, 'A Note on Investment', in Tabitha Moses and Jon Barraclough, *Tabitha Moses: Investment* (self-published, 2014).

18. C. Lewis, cited in K. Bill, *Plug in for Life: The Remarkable Story of Moreen Lewis, Her Faith and Triumph Over Kidney Failure and the Tragedy of Britain's Kidney Machine Crisis* (Oliphants, 1968), p. 96.

19. T. M. Boon, 'Histories, Exhibitions, Collections: Reflexions on the Language of Medical Curatorship at the Science Museum after Health Matters', in R. Bud, B. S. Finn and H. Trischler (eds), *Manifesting Medicine: Bodies and Machines*, vol. 1 (Science Museum, 1999), p. 124.

20. A. Gawande, *Complications: A Surgeon's Notes on an Imperfect Science* (Profile, 2007), p. 229.

FURTHER READING

Alberti, S. *Morbid Curiosities: Medical Museums in Nineteenth-Century Britain* (Oxford University Press, 2011).

Anderson, S. C. (ed.). *Making Medicines* (Pharmaceutical Press, 2005).

Arnold, K., and D. Olson (eds). *Medicine Man: The Forgotten Museum of Henry Wellcome* (British Museum Press, 2003).

Barnett, R. *The Sick Rose Or; Disease and the Art of Medical Illustration* (Thames & Hudson, 2014).

Berkowitz, C. *Charles Bell and the Anatomy of Reform* (University of Chicago Press, 2015).

Bourke, J. *Dismembering the Male: Men's Bodies, Britain and the Great War* (Reaktion, 1999).

— *The Story of Pain: From Prayer to Painkillers* (Oxford University Press, 2014).

Bud, R., B. S. Finn and H. Trischler (eds). *Manifesting Medicine: Bodies and Machines*, vol. 1 (Science Museum, 1999).

Bynum, W. F., and H. Bynum. *Great Discoveries in Medicine* (Thames & Hudson, 2011).

Bynum, W. F., and R. Porter (eds). *Companion Encyclopaedia of the History of Medicine* (Taylor and Francis, 1997).

Chadarevian, S. de, *Designs for Life: Molecular Biology after World War II* (Cambridge University Press, 2002).

Chadarevian, S. de and N. Hopwood. *Models: The Third Dimension of Science* (Stanford University Press, 2004).

Chaplin, S. *John Hunter and the 'Museum Oeconomy' 1750–1800* (unpublished PhD thesis, Kings College London, 2009).

Church, R., and E. M. Tansey. *Burroughs Wellcome & Co.: Knowledge, Trust, Profit and the Transformation of the British Pharmaceutical Industry, 1880–1940* (Crucible, 2007).

Crellin, J. K., and J. R. Scott. *Glass and British Pharmacy 1600–1900: A Survey and Guide to the Wellcome Collection of British Glass* (Wellcome Institute of the History of Medicine, 1972).

Cunningham, A. *The Anatomist Anatomis'd: An Experimental Discipline in Enlightenment Europe* (Ashgate, 2010).

Daston, L. (ed.). *Things that Talk: Object Lessons from Art and Science* (Zone, 2007).

Davidson, R., and L. A. Hall (eds). *Sex, Sin and Suffering: Venereal Disease and European Society Since 1870* (Routledge, 2001).

Draycott, J., and E. Graham (eds). *Bodies of Evidence: Ancient Anatomical Votives Past, Present and Future* (Routledge, 2017).

Francis, D. (ed.). *Faith and Transformation: Votive Offerings and Amulets from the Alexander Girard Collection* (Museum of New Mexico Press, 2007).

Gawande, A. *Complications: A Surgeon's Notes on an Imperfect Science* (Profile, 2007).

Hasegawa, G. R. *Mending Broken Soldiers: The Union and Confederate Programs to Supply Artificial Limbs* (Southern Illinois University Press, 2012).

Hill, J. M. *Cultures and Networks of Collecting: Henry Wellcome's Collection* (unpublished PhD thesis, Royal Holloway, University of London, 2004).

Hill Curth, L. (ed.). *From Physick to Pharmacology: Five Hundred Years of British Drug Retailing* (Ashgate, 2006).

Holloway, S. W. F. *Royal Pharmaceutical Society of Great Britain 1841–1991: A Political and Social History* (Pharmaceutical Press, 1991).

Homan, P. G., B. Hudson and R. C. Rowe. *Popular Medicines: An Illustrated History* (Pharmaceutical Press, 2007).

Jackson, L. *From Atoms to Patterns: Crystal Structure Design from the 1951 Festival of Britain* (Wellcome Collection, 2008).

Jackson, R. *Doctors and Diseases in the Roman Empire* (British Museum Press, 1988).

Jordanova, L. *Sexual Visions: Images of Gender in Science and Medicine between the Eighteenth and Twentieth Centuries* (University of Wisconsin Press, 1993).

Kirkup, J. *A History of Limb Amputation* (Springer, 2007).

Larson, F. *An Infinity of Things: How Sir Henry Wellcome Collected the World* (Oxford University Press, 2009).

Lowy, I. *A Woman's Disease: The History of Cervical Cancer* (Oxford University Press, 2011).

Lupton, D. *Medicine as Culture: Illness, Disease and the Body* (Sage, 2012).

Maerker, A. *Model Experts: Wax Anatomies and Enlightenment in Florence and Vienna, 1775–1815* (Manchester University Press, 2011).

Marchant, J. *Cure: A Journey into the Science of Mind Over Body* (Broadway, 2016).

McCallum, J. E. *Military Medicine from Ancient Times to the 21st Century* (ABC-CLIO, 2008).

Mihm, S., K. Ott and D. Serlin (eds). *Artificial Parts, Practical Lives: Modern Histories of Prosthetics* (New York University Press, 2002).

Mitchell, P. (ed.). *Anatomical Dissection in Enlightenment England and Beyond: Autopsy, Pathology and Display* (Ashgate, 2012).

Moerman, D. E. *Meaning, Medicine and the 'Placebo Effect'* (Cambridge University Press, 2002).

Moscucci, O. *The Science of Woman: Gynaecology and Gender in England, 1800–1929* (Cambridge University Press, 1990).

Turner, H. *Wellcome: The Man, the Collection, the Legacy* (Heinemann Educational, 2001).

Packard, R. M. *A History of Global Health: Interventions into the Lives of Other Peoples* (Johns Hopkins University Press, 2016).

Phillips, G. *Best Foot Forward: Chas. A. Blatchford & Sons, Ltd. Artificial Limb Specialists 1890–1990* (Granta Editions, 1990).

Porter, R. *Blood and Guts: A Short History of Medicine* (Allen Lane, 2002).

— *Bodies Politic: Disease, Death and Doctors in Britain, 1650–1900* (Cornell University Press, 2001).

— *The Greatest Benefit to Mankind: A Medical History of Humanity* (Norton, 1998).

Richardson, R. *Death, Dissection and the Destitute* (Phoenix Press, 2001, 2nd edn).

Russell, G. 'The Wellcome Historical Medical Museum's Dispersal of Non-Medical Material 1936–1983', *Museums Journal Supplement*, vol. 86 (supplement), 1986.

Spandler, H. *Asylum to Action: Paddington Day Hospital, Therapeutic Communities and Beyond* (Jessica Kingsley, 2006).

Spongberg, M. *Feminizing Venereal Disease: The Body of the Prostitute in Nineteenth-Century Medical Discourse* (Macmillan Press, 1997).

Symons, J. *Buller's Campaign* (House of Stratus, 2008).

Warner, J. H., and J. M. Edmonson. *Dissection: Photographs of a Rite of Passage in American Medicine, 1880–1930* (Blast, 2009).

Williams, G. *Paralysed with Fear: The Story of Polio* (Palgrave, 2015).

Withey, A. *Physick and the Family: Health, Medicine and Care in Wales, 1600–1750* (Manchester University Press, 2011).

Wright, J. *A History of War Surgery* (Amberley, 2011).

AUTHOR BIOGRAPHIES

MURIEL BAILLY is Assistant Curator of Medicine at the Science Museum. She is interested in the interpretation and display of molecular models and, more generally, in finding innovative ways to convey complex scientific topics to a wide audience. Prior to joining the Science Museum, she worked as an Exhibition Assistant at Wellcome, where she led the contemporary science programme, 'Medicine, What Now?'.

SARAH BOND is Associate Curator of Medicine at the Science Museum. She has led on two of the new galleries showcasing the diversity of the Medicine collection and revealing the human experience of health and wellbeing. Her research interests relate to the intersections between memory and material culture. Prior to joining the Science Museum in 2014, she was Assistant Curator of Moving Image and Sound and Visitor Experience Assistant at Wellcome.

DR JACK DAVIES completed his PhD at the University of Kent in 2017. His thesis examined the use of stately homes as hospitals during the First World War. During this he worked as the Resident Historian at Turner Contemporary, as well as holding a student research position at the University of California, Berkeley. Since completing his PhD, he has worked as Assistant Curator on the new Medicine Galleries at the Science Museum.

STEWART EMMENS is Curator of Community Health at the Science Museum. He has worked in the Museum for many years on a wide range of medical exhibitions and web resources, as well as curating a number of medical collections. His main research interests are limb prostheses, military medicine and urban public health. He was the lead curator on the Science Museum's First World War centenary exhibition, *Wounded: Casualty, Conflict and Care*, and led on one of the new Medicine Galleries focusing on public health and the challenges of caring for communities and populations.

BRIONY HUDSON is a freelance museum curator, pharmacy historian and lecturer. Having worked in-house at museums including the Royal Pharmaceutical Society and Royal College of Surgeons, her freelance work has included exhibition, policy and publication projects for a number of the medical royal colleges, the Wellcome Library and pharmacy schools. She is past President of the British Society for the History of Pharmacy, past Chair of London Museums of Health and Medicine and current Deputy President of the Faculty for the History and Philosophy of Medicine and Pharmacy at the Society of Apothecaries.

SELINA HURLEY is Curator of Medicine at the Science Museum. She has led on one of the new Medicine Galleries, focusing on her interests in therapeutics including surgery and *materia medica*. Since joining the Museum in 2006, she has worked on the Science Museum's web project, *Brought to Life*, which started a fascination with medical collections. In addition, she was lead curator for *The Chronophage: A Time Eating Clock* in 2011 and *Climate Changing Stories* in 2012, and curated the contemporary science element of *Mind Maps: Stories from Psychology* in 2013.

DR ANNA MAERKER is a medical historian and Senior Lecturer in History of Medicine at King's College London. Her work investigates the historical emergence of modern science and medicine as authoritative practices of knowledge production through a focus on material culture and everyday practices. She is the author of *Model Experts: Wax Anatomies and Enlightenment in Florence and Vienna, 1775–1815* (Manchester University Press, 2011). At the Max Planck Institute, she co-curated the exhibition *Objects in Transition*, and she also contributed to the web project *Brought to Life* at the Science Museum. She is a founding member of the Public History Seminar at the Institute of Historical Research, London, and now serves as Historian in Residence at the Hunterian Museum of the Royal College of Surgeons.

NATASHA McENROE is the Keeper of Medicine at the Science Museum. Her previous posts include Director of the Florence Nightingale Museum, Museum Manager of the Grant Museum of Zoology and Comparative Anatomy and Curator of the Galton Collection at University College London. From 1997 to 2007, she was Curator of Dr Johnson's House in London's Fleet Street and has also worked for the National Trust and the Victoria and Albert Museum. Natasha was co-editor of *The Hospital in the Oatfield: The Art of Nursing in the First World War* (Strange Attractor Press, 2014).

EMMA STIRLING-MIDDLETON works on the social history and anthropology of science, technology and medicine. Before joining the Science Museum in 2017, Emma worked at the international touring exhibitions company Nomad Exhibitions, where she developed blockbuster exhibitions in partnership with major cultural institutions across ten different countries. Emma is currently a Project Curator at the Science Museum, researching Chinese science, technology and medicine, and she develops exhibitions in partnership with the Palace Museum in Beijing and the Hong Kong Science Museum.

DR OISÍN WALL is an historian and curator at the Centre for the History of Medicine in Ireland at University College Dublin. His current research is focused on prisoners' protests and their health in the 1970s and 1980s. He published his first book, *The British Anti-Psychiatrists: From Institutional Psychiatry to the Counter-Culture, 1960–1971*, in 2017. He previously worked at the Science Museum where he curated *Journeys Through Medicine*, and spent three years working on one of the new Medicine Galleries at the Science Museum.

INDEX

Illustrations are in *italics*.

Abraham, Edward 188
acquisitions *see* collecting
Action Medical Research 92
Acton, William 128
Adventures of Roderick Random,
  The (Smollett) *165*
advertisements
    apothecaries and pharmacies
      155, *161*, *162*, *163*, *167*, *170*,
      171, *173–6*
    Burton, Elizabeth *58*
    medical appliances and apparatus
      21, *137*
    medicines *170*
airships 15
Akron airship 15
Alexander Girard Collection 216
Alphonso XIII of Spain 22
Alternative Limb Project 62–3
alternative medicine 222
America airship 15
American Cancer Society 135
American Journal of Obstetrics
  and Gynecology 135
amino acids 186
amputations
    current reasons for 61
    employment 51, 58, 60
    splints 117
    surgical history 41
    tuberculosis 41
    war 41, 43, 47, 48, 101, 114–16,
      117–18
amulets 30, 202, 203, 214–15, 216,
  217, 220, 221
anaesthetics 41, 99, *100*, 101
anatomical Venuses 180, *182–3*
anatomy
    digital technology 79, *80*
    humours theory 177–8
    models 69, 70–3, 74–7, 177–8,
      180, *182–5*, 186, 188–90, *191*,
      195–8, 199, 200
    scientific observation 178
    see also dissections
Anatomy Act 73–8, 185–6
Anderson, Stuart 173
animals
    apothecaries 160
    dissection 67, 177–8
    medicines 170
    research 132
antibiotics 41, 122–3, 187, 198
anticoagulants 118
antipsychotics 150, *150*
antisepsis 19–20, 21, 41, 99, *100*, 119
apothecaries *154*, 155, 157–62, *158*,
  159, 162

Apothecaries Act 166
Apothecaries shop (Tregaskis) *168*
Archaeological Museum of Ancient
  Corinth 207, *209*
arms
    prosthetics 43–5, 56–8, *57*, *61*
    splints 94, *95*
Army, British 108, 114
Army Nursing Service 108, 114
Arnold, John 160
arteries 26, 27, *178*
artificial limbs *see* prosthetics
Ashanti people 216, *218–19*
ashtrays 35
asylums 144, 145, 148–9, 152
athletics 64, 65
Atlas of Exfoliative Cytology
  (Papanicolaou) 135
Atwood, Margaret 199
auctions 26, 29, 30
autographs 26
autopsies 67
Auzoux, Louis 190, *190–1*, 195,
  195–7

baby carriers 16, *16*, *17*
bacteria
    antibiotics 41, 122–3, 187, 198
    germ theory 99
    superbugs 198
Baden-Powell, Robert 216
Balaclava 108
balloons, military 110, *111–13*
barber-surgeons 99
Barrie, J. M. 39
Baskerville, Robert 157, 160
Baynam, Tracey 32, *33–4*
BeBionic hand 61–4, *64*
Beken and Son *171*
Bell, Jacob 155
Bell, John *154*, 155
Benton, Claude *174*
Beral, Valerie 139
bile 177
Bill, Keith 224
birdcages 151, *152*
Black Notley Hospital, Essex 90
Blackman, Winifred 27
Blackstone, Mrs 95
Blackwood's Magazine 166
Blondin, Charles 30
blood, humours theory 177
blood circulation 178
blood loss 41
blood transfusions 41, 117–18, *119*
blood types 117–18
Blue, Ruth 34
'Body Alteration' series (Quinn) 199
body imprints on artefacts 204, 225
body-snatching 70, 73, 78, 185

Body Worlds exhibition *198*, *199*
Boer War, Second 108–14
bones
    polio 86
    splints 116–17, *118*, 124, *125*
Boots 172
Borden, Mary 114
Borer, Mary Irene Cathcart 27–9,
  27, *28*
Boston Children's Hospital 87
Both, Edward 87
braces 86
Bragg, Laurence 180
Bragg, William 180
brain, effect of mercury 131
breathing 83, 86, 89–90, 95
Brighton County Borough Lunatic
  Asylum 148–9, *151*, *152*
    see also St Francis Hospital,
      Sussex
British and Colonial Pharmacist
  172
British Pharmaceutical Conference
  173
British Polio Fellowship 91
Brown, Andrew 19, 20, 21
Brumwell, Marcus 199
Buddhism 209
bulbar paralysis 86–9
Burke, William 73, 78, 185
burns 123
Burroughs, Silas 13
Burroughs Wellcome & Co. 13–15,
  22
Burton, Elizabeth 56–8, *58*
Byrne, Charles 70, *70*

C-A-T tourniquet 124, *124*, *125*
Cairo 23
Call the Midwife (TV show) 34
callipers 83, 86, 92, *93*, 95
Cambodia 52, *54–5*
Cambridge, Theatre of Anatomy *81*
cancer
    cervical 127, 131–3, 135–40,
      142–3
    genetic research 198
    oral 140
    radiotherapy 225, 226
candlesticks 20, *20*
capsules 173
Captain Hook 39, 41
carbolic sprays 19–20, 99, *100*
carboys 163, *164–5*, 166, *171*, *173*,
  174
'care in the community' 151, *152*
Carnes artificial arms 42, 43–5, 64
Carnes Artificial Limb Company
  43–5
Carrel, Alexis 116

Carrel-Dakin method 116, *117*
Casper, Monica J. 136
Cassel Hospital, London 145
casts 86, 90, 91, 96, 185
cataloguing 27, 28, 30, 205
cauterisation 41
cell theory 132
Celox 124, *124*, *125*
Cerletti, Ugo 145
cervical cancer 127, 131–3, 135–40,
  142–3
Chailey Heritage, Sussex 90–1
Chain, Ernst 188
Changi Prison, Singapore 52, *53*
charcoal water filters 108, *109*
charities 91–5, 107, 114, 224
charms 203, 216–20, *218–19*
chemical convulsive therapy 145,
  149
Chemist and Druggist 172
Chicago
    1893 World Fair 216
    Century of Progress Exposition
      15
childbirth 70, 73, 209, *211*
children
    amulets 216, *217*, 220, *221*
    fundraising 92
    polio 83, 84–6, 89, 90, 95, 97
    thalidomide 33
    trauma 85–6, 97
chitosan *124*
chlamydia 175
chloroform 101
Christianity 209–11, *212–13*
cinchona bark 22, 24, 157, 166
Clark, Catherine 73
Clarke, Adele E. 136
clotting 117–18, *124*
Coene, Constantin Fidèle 48
Coldwell, Arthur 159
Coleman, Dulcie 137
collecting
    agents 16–19, *20–3*, 24, 27
    auctions 26, 29
    'comparative method' 203
    duplications 26
    MacCulloch, William Mansell 24
    material culture 151–2
    pharmaceutical business 36
    pharmacies 155
    records *17*, 30 *see also*
      cataloguing
    relationships 125
    Science Museum 30–3, *34–6*
    Wellcome, Henry 10, 13, 15–19,
      20–2, 36, 99, 203–4, 205
collection boxes 92–5, *92*, *93*
College of Physicians 68, 166
comas 145, 147, *147*

muscles, polio 86
museums
    Archaeological Museum of
        Ancient Corinth 207, 209
    dissections 68, 70
    La Specola, Florence 180
    Museum of International Folk
        Art, Santa Fe 216
    wax cabinets 73
Myers, Oliver Humphrys 29
myoelectric technology 61, 64
myoglobin 179, 190, 192–3, 199

Nairn, Peter 130
narrative see testimonies
National Cervical Cancer Prevention
    Campaign 136
National Cervical Cytology
    Screening Service 136, 137, 138
National Foundation for Infantile
    Paralysis (NFIP) 90, 91, 92
National Health Insurance Act
    171–2
National Health Service (NHS)
    123–4, 136, 138, 172, 224
National Kidney Centre 224
National Pharmaceutical Association
    173
Native Americans 16, 220, 221
necklaces 216, 217
Needham, Thomas 159
nervous system, mercury 131
netsuke 36, 37
neurons, polio 83
Nightingale, Florence 102–7, 108
Nobel Prize 135, 139, 188, 198, 199
noses
    reconstruction 118–19
    reconstructions 123
Nosso Senhor do Bonfim, Brazil 213
Nuffield Report 175
nursing
    education 107
    military 102, 104–6, 107–8, 114
    polio 85, 87, 89, 90, 95
nux vomica 157, 166

obstetric phantoms 70, 73
occupational therapy 51, 119
'officer's arm' 45
opium 166
oral cancer 140
Ottobock Healthcare, running
    blades 65

padded cells 144, 145
Paget, Henry, Marquess of Anglesey
    47–8, 48
pain 204, 207
panaceas 147

Papanicolaou, George N. 132–5,
    133–4, 143
Paralympics 64
parrots 151
Pasteur, Louis 22, 99, 101
patents 162, 170
Paterson, Arthur Spencer 145–7,
    146, 151
pathology specimens 185, 186, 200,
    204
patient care, communication 85–6,
    97
patient experiences 225–7
patient record holders 20, 20
patient testimonies
    collecting 224, 225
    interpretation 85, 89, 90, 95–7,
        227
    mental health 151–2
Pavón, Antonio José 22
Pearson, Drew 89
peg-legs 38, 39, 41, 41, 51, 52, 54–5
penicillin 122–3, 123, 187, 188,
    195, 198, 199
Peninsular Wars 99
Pepys, Elizabeth 157
Pepys, Samuel 26, 157
Peru 22, 22
Perutz, Max 190, 194, 195
pestles 12, 36, 37
'phantom limb pain,' limbs 61
Pharmaceutical Journal 166
Pharmaceutical Society of Great
    Britain (PSGB) 155, 167, 172
pharmacies
    apothecaries 154, 155, 157–62,
        158, 159
    Arabian 22–4, 25
    Bell, John 154
    chemists and druggists 162–8
    collecting 155
    development of imported drugs
        157
    dispensing 168–70, 172, 173, 175
    diversification 171, 172
    home remedies 157
    legislation 166, 168
    medical training 166–7
    Nuffield Report 175
    professionalisation 166–7, 170–1,
        173, 175
    reconstructions 156, 157
    social attitudes 155
pharmacy jars
    Arabian 23, 23, 24
    collecting 24–6, 26
    decoration 159–60, 160, 167, 169
    leeches 169
    move to glass 163, 168
    specie 163–6, 164–5

Phillips, Van 64
Philpots, Richard 157
phlegm 177
photography 171
pianists 56–8, 57
Pitt Rivers, Augustus 203
placebo effect 204
Plas Newydd, Anglesey 48
plastic surgery 118–19, 123
plastination 199
Plug in For Life (Bill) 224
pluralism 222
poisons, glassware design 168
polio
    contagion 83–4
    epidemics 84
    films 91
    fundraising 91–2
    patient experiences 84–6
    respiratory paralysis 83, 86–90,
        92
    symptoms 83
    vaccine 83, 84, 91, 95
politics, sexuality 142
Pompeii 126, 127
Poor Laws 167
post-traumatic stress disorder
    (PTSD) 145
power failures 87–9
prayer 207, 216, 222
pregnancy 33, 209, 211, 220
Prescott, Thomas 160
prescriptions 155, 172, 175
Price, Robert E. 170
prisoners of war 52, 53
professionalisation 99, 108, 167,
    170–1, 173, 175
Prometheus Medical 125
Prontosil 122
prosthetics
    appearance 60, 61, 61, 62–3,
        64–5
    collection images 32, 38, 40, 43,
        46–7, 49, 53–5
    comfort 33, 48
    complexity 43–5, 64
    costs 45, 64
    earliest in collection 41–3
    emotions 204
    employment 51
    home-made 38, 39, 51–2
    home repairs 48, 49
    'Jaipur limb' 52–6, 56
    literature 39–41
    manufacture 43–5, 47, 51, 52,
        53, 58
    military 56, 125
    mind control 61, 64
    'pylon legs' 48–51, 50
    running blades 64, 65

social attitudes 39
split hook attachment 45, 45
technology 60–1, 64, 65
thalidomide 32, 33–4
tools 51, 52, 60
prostitution 122, 128, 142
proteins 190, 192–4
psychiatry
    antipsychotics 150
    'care in the community' 151, 152
    early treatments 145–7
    fall in number of inpatients 151
    group therapies 150
    hospitals 144
    legislation 150
    reform campaigns 151
    treatment criticisms 149
PTSD (post-traumatic stress
    disorder) 145
public health campaigns
    pharmacies 173–6
    screening programs 127, 132,
        142–3
    smoking ban 34–5, 34
    venereal diseases (VD) 141
'pylon legs' 48–51, 50
pyrotherapy 145

Queen Alexandra's Imperial Nursing
    Service 114
Queen Mary's Hospital, Roehampton
    33, 43, 43, 51, 52, 56
Quincy, John 157
quinine 22, 157, 166
Quinn, Marc 199

rabbits 36, 37
Rackstrow, Benjamin 73
Rackstrow's Museum 73
radiotherapy 225, 226
Récamier, Joseph 127–8
recipe books 157
reconstructions 23–4, 25, 30, 31,
    156, 157
Redgrave, Michael 91
Rees, T. P. 150–1
registration 27–9
rehabilitation 119, 123
Reinhardt, G. B. 171
relationships 125, 220, 225
religion 30, 209–11, 227
religious orders, provision of medical
    care 24
reliquaries 66, 67
Renou, Jean de 159
replicas 20, 41–3, 42
residential homes 175
respirators 82, 84, 85, 86–90, 87,
    88, 95, 97
respiratory paralysis 83, 86–9, 92

ACKNOWLEDGEMENTS

The authors would like to thank the following people for their help with the production of this book. Wendy Burford and Charlotte Grieveson for their editorial expertise, ably assisted by Karolina Koziel, Tig Thomas and Anjali Bulley. Kevin Percival, Kira Zumkley and Jennie Hills were responsible for the beautiful photography created especially for this publication. We are also grateful to Laura Lappin at Scala and Peter Dawson and Alice Kennedy-Owen at Grade Design.

Thanks are due to our hard-working colleagues in the Science Museum's Conservation and Registry departments who care for the collections, and whose work is largely unseen but without whom the Science Museum could not function. We are grateful for the ongoing support and friendship of our colleagues at Wellcome Collection, and the London Museums of Health and Medicine and for the wise guidance from the Medicine Gallery project's Advisory Panel.

We are especially grateful to the following people: Katy Barrett, Frances Benton, John Betts, Tim Boon, Jessica Bradford, Robert Bud, Keeley Carter, Imogen Clarke, Katie Dabin, Anna Dejean, Jennifer Francis, Taragh Godfrey, Christine Gowing, Matthew Johnston, Joe Kenway, Isabelle Lawrence, Karen Livingstone, Jack Mitchell, Vandana Patel, Emily Scott-Dearing, Roisin Tierney, Lucy Trench and Margaux Wong. Members of the Thalidomide Society have been very helpful, and we thank them for their input, in particular Secretary Ruth Blue.

Finally, we would like to thank Tilly Blyth, Alison Boyle, David Rooney and all the curators and researchers at the Science Museum, past and present. Their knowledge and dedication to the Science Museum collection is truly extraordinary.